Resilience

# Introduction

Landscapes are constantly changing and evolving. From rural settings to cities, things appear, disappear or are replaced on a daily basis. But traces of the past are all around us: paths, buildings and hedges all betray the presence of former populations. Taking a bird's eye view, using planimetric documents (maps, topological surveys, etc.) in conjunction with modern aerial photography, the coexistence of old and new is plain to see. The layout, or ground footprint, of elements in a landscape (buildings, paths, hedges and ditches, fields, gardens, and so on) tends to subsist for far longer than the most apparent forms, which we encounter in our daily lives. In what follows, the terms "shape" and "ground footprint" will be used to distinguish between these two layers. The "**shape**" of a landscape is that which we see directly. This term, borrowed from the field of topography, is used in archeogeography to refer to all types of forms in three-dimensional space (i.e. "volumes"), be they man-made or of natural origin (Chouquer 1997a, p. 15). The space which the volume occupies on the ground, that is, its footprint, is a two-dimensional form, seen from above, which can be mapped. In archeogeography, the ground footprint of elements in a landscape has been referred to as the *tracé en plan* (map trace; Chouquer 1997a, p. 15), ***forme en plan*** (plan form) or ***tracé*** (trace; Robert 2003a, p. 117 and 127).

In the context of this work, the word "landscape" is used essentially to refer to the material dimension, but territorial boundaries enshrined in law (land boundaries and territorial district boundaries) are also taken into account. Although these limits are not necessarily visible on the ground, they are clearly shown on maps and land registers, and are remarkably persistent relics of earlier forms of landscape organization. While the shape of town districts, land holdings, roads, drainage and irrigation networks, habitat distribution patterns, etc., has undergone dramatic changes over time, these elements are rooted in earlier, persistent ground footprints and divisions. Landscape is like the ship of Theseus, kept in harbor by the Athenians, who progressively replaced each plank as it rotted until the whole ship had been replaced.

Elements in a landscape undergo constant renewal, while retaining, to some extent, the identity of the earlier landscape. This interweaving and internesting of past and present, bearing the marks of successive generations of change, raises questions concerning the coexistence of these multiple forms.

Researchers have studied landscape evolutions, by means of morphological analysis, since the 19th century. In Part 1 of this book, we shall consider the way in which authors have apprehended the dynamics of shape and have conceptualized persistence and change in the forms of a landscape. Different approaches have been used in specific contexts of spatial and social change; morphological analysis, as a scientific approach to the study of landscape, has moved from an essentially cognitive to an essentially normative approach and repeated over time. For this reason, the approaches taken by researchers interested in the "temporal" aspects of landscape change (archeologists, geographers, historians, etc.) are presented alongside the approaches taken by those involved in shaping our current habitat (architects, urban planners, etc.). In this way, we aim to highlight the articulations and distinctions between temporal and spatial aspects in architecture and urban planning. Leaving behind the traditional approach, in which the continuity of forms in a landscape is seen in terms of inertia or as a form of palimpsest, we shall show how the notions of complex systems, self-organization and resilience provide a new and effective framework for morphological landscape analysis. Particular attention will be paid to the concept of ecological resilience and its application to the field of archeogeography, notably with respect to the multiple temporalities at work in landscape systems.

Part 2 concerns the emergence and development of the conceptual framework of resilience, particularly in the context of ecology. Research into resilience first began in the 1970s, following on from the adoption of the complex system and self-organization paradigms; this concept provides a way of accounting for temporal dynamics and allows us to address interactions across multiple levels in the organization of complex systems. Initially limited to the fields of physics, psychology and ecology, the concept of resilience only really took off in the late 2000s, in the context of campaigns to draw attention to the need for action to combat climate change and manage catastrophes. Moving out of the purely cognitive field, the term "resilience" has taken on multiple meanings and multiple implications. In this book, we shall focus on the conceptual framework of "ecological" resistance, as developed by the Canadian ecologist Crawford S. Holling and promoted by the Resilience Alliance network. Ecological resilience is a powerful heuristic concept with the ability to change our perception of the way in which interactions between societies and milieux evolve. We thus present the concept of resilience as a property of social-ecological systems, before describing the ways in which it has been used in analyzing spatial systems and landscapes in the fields of archeology, geography and archeogeography. Particular attention will be given to the place assigned to heritage

# Introduction to Part 1

Morphological analysis, the science or study of forms, first emerged in the 19th century. First used in biology, with respect to the external form and structure of living beings, and in linguistics, with respect to the different forms and rules governing the internal structure of words[1], it is later used in the social sciences to describe functional structures and groups. In geography, history and architecture, the development of new representations of landscape, using planimetry, gave rise to a school of research based on morphological analysis. This approach was used across the domains of human geography, history and archeology, based on the shared idea that material or spatial realizations – or artifacts – cast light on social organizations.

The term **form**, derived from the Latin *forma*, which refers to both the "mold" and the "molded object", concerns both the visible appearance of an object and that which gives it its shape. In this sense, it also relates to a "model to imitate" and may be understood as an "organization in accordance with a norm" (Rey 1993, p. 814). Form, thus, relates both to the appearance of organizations and to the organizing principle that underpins them (Chesneau and Roncayolo 2011, p. 156). The same ambiguity with respect to knowledge and ontology is also found in the word **morphology**, a term coined by Goethe in 1790. Originally used to refer to the study of the external configuration of an organ or living being, the term came to be used to refer to the actual form of living organisms (Rey 1993, p. 1275). Studying the prevailing practices in morphological analysis can tell us much about the way a society perceives a landscape and understands the way in which it marks this landscape, and consequently about the relationship between the society and its environment; the landscape is the result of this relationship. The nature of the relationship varies depending on the historical context, but also on the media used to perceive the landscape. Developments such as topographic cartography, in the late

1 https://www.cnrtl.fr/definition/morphologie, accessed July 24, 2020.

19th century, aerial photography, in the wake of the First World War, and widespread archeological campaigns from the 1990s onwards all played a decisive role in the way in which researchers perceive and understand landscape forms.

The earliest studies of landscape forms adopted a diachronic approach, considering past and present in parallel; in the decades following the Second World War, however, a synchronic approach came to dominate the field. This approach is built around a strata-based framework where the present buries, and effectively cancels out, the past. Observations made in the course of rescue archaeology in the 1990s challenged this stratified vision of landscapes, highlighting the true complexity of temporalities, while increasingly strong cross-disciplinary connections between historians, archeologists and environmental specialists changed perceptions of the relationships between societies and environments. These developments formed the backdrop for a resurgence of interest in morphological analysis of historical landscapes, culminating in the emergence of a whole new discipline: archeogeography[2]. This approach promotes a vision of landscape in which the links between man and milieu play a key role, moving beyond the simple nature/culture division. In this first part, we shall consider the way in which past authors, from the late 19th century onward, envisioned the relationship between space and time in a landscape. We shall also explain why modern archeogeographers must move beyond the morphological analysis methods developed during this earlier period, demonstrating the need for a new and innovative theoretical framework.

2 The French term is *archéogéographie*; in English, the term "landscape archeology" is also widespread.

# 1

# Landscape: The Resistance of the Past?

## 1.1. The past in the present

### 1.1.1. *Architectural and morphological persistences*

Humanity has long been aware that certain constructions persist beyond the societies that shaped them. As early as the early Middle Ages, printed works and iconographic representations highlighted the presence of ancient constructions in the urban fabric of the city of Rome (D'Amico 2009). These constructions are presented as anchoring elements, ensuring that the city itself remained "eternal" (Djament-Tran 2011). Roads were also identified as permanent features of the landscape (D'Urban 1837, p. 415), and their continued presence is reflected in place names. The persistence of these roads was perceived locally by users well before Nicolas Bergier, a magistrate, made his first theoretical proposals in the early 17th century concerning the conditions under which these great ancient routes had survived the centuries (Bergier 1622).

By the late 19th century, scholars were well aware of the persistence of other, more extended and complex constructions in the landscape: town and city layouts, field patterns, village ground footprints and settlement patterns often followed agrarian structures established in the Middle Ages or even earlier. These elements attracted the attention of historians and legal specialists, philologists, geographers and architects from across the globe. This shared interest is partially explained by the development of increasingly detailed representations of space. Cartography made a significant leap forward at this time due to an increase in the precision of geodesic measurements and to new projection systems. Vast mapping campaigns were carried out over the course of the 19th century, establishing or updating national cartographic references by means of detailed surveys, at scales of 1:10,000, 1:40,000, etc. At the same time, cadastral surveys were carried out on an even larger scale: 1:1,250, 1:2,500, etc. (Steinberg 1982; Maurin 1992). Finally, the increase in the production and availability of travel guides provided further helpful tools for the first city and town planners, who were able to consult and compare existing layouts (Lavedan 1926a, p. 91).

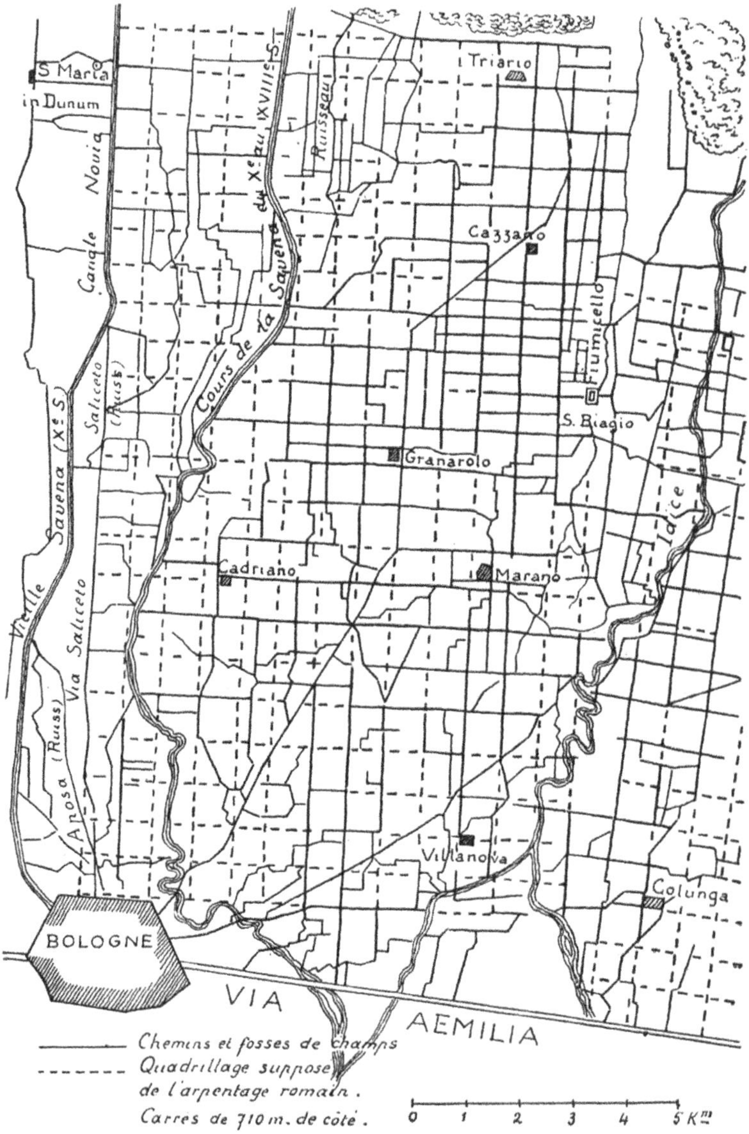


**Figure 1.1.** *Reproduction of an ancient centuriation plan near Bologna, studied by the archaeologist Alfonso Rubbiani, from the 19th-century cartography in Albert Grenier's Manuel d'archéologie gallo-romaine (Grenier 1934, p. 17)*

This new documentation clearly showed the continued existence of ancient forms in the landscape. Concerning ancient planned plots in Campania, for example, the French archaeologist Albert Grenier indicated that the land "appeared to be" divided into square plots with a "very clear grid pattern". He added that the most striking example was found in northern Italy (Figure 1.1) (Grenier 1934, p. 15). From the 19th century on, philological studies of texts relating to ancient surveying practices, whereby certain territories were divided along rectilinear lines intersecting at right angles into parcels or plots known as "limitations" or "centuriations" (*centuriae*),[1] were compared with material traces. Using cross-analysis of topographic maps, cadastral records, aerial photographs and field observations in Italy, France, North Africa, the Near East, etc., historians attempted to reconcile the material traces with the texts (Chouquer 2008a).

### 1.1.2. *Looking to the present to uncover the past: regressive history*

In countries bearing fewer marks of Roman occupation, many researchers focused on medieval agrarian landscape structures. In 1895, August Meitzen, a professor of statistics and economics at the University of Berlin, posited that a type of land division that he had observed on cadastral plans was, in fact, the imprint of legal land plots established in late antiquity and the early Middle Ages (Meitzen 1895). Meitzen's analytical method was widely disseminated, and historians began to pay increasing attention to cadastral plans as source materials. Inspired by Meitzen's work, F.W. Maitland (1850–1906), professor of law at the University of Cambridge, began to study the origin of the grouped villages in the open field system and the dispersal of the English bocage. He stated: "Two little fragments of the original one-inch ordnance map will be more eloquent than would be many paragraphs of written discourse" (Maitland 1987, p. 16). Maitland's work gave rise to a new tradition of research in historical topography in Great Britain, first based on map analysis, and later on aerial photography (Darby and Williams 2002, p.18).

In France, the historian Marc Bloch (1886–1944), familiar with the work then coming out of Germany and Great Britain, formalized the "inverse method" in 1931. Bloch's approach consisted of "reading history backwards" from texts and cartographic representations produced during the 18th century, a period in which landscapes and agrarian practices began to be better documented (Bloch 1988, p. 49). According to the historian Adriaan Verhulst, the regressive method consists of:

1 When these cuttings present "grids of equidistant axes, defining square or rectangular units, called centuriae" (see entry "centuriatio" in the Dictionary of Surveying on www.archeogeographie.org, accessed on January 20, 2018).

> [Beginning with] the least unknown, which is usually also the most recent or closest to us, such as the present landscape or the nineteenth-century cadastral plan, in order to travel backwards into the past by means of clues which become increasingly difficult to interpret the further back we go, but which occur in a clear historical sequence. (Verhulst 1995, p. 20)[2]

Recent documentation is used for reasons of simple necessity, where no historical equivalent exists for the period in question. For urban historians, planimetric documents are the only practical source for obtaining ancient city plans, since it is impossible to excavate a whole town or city. For Marcel Poëte, one of the fathers of urban morphology in France (alongside Pierre Lavedan)[3], there are clear and significant "medieval legacies" in the current layouts of major cities (Poëte 1924, p. 7). Ordnance maps, cadastral documents and modern plans are thus "perfectly acceptable" source documents for the study of medieval cities, and even for the ancient period (Lavedan 1926a, p. 94). Lavedan generalized these into "a principle, which, if not universal and absolute [...], is at least applicable to the majority of cases: the *rule of persistence of the plan*"[4], which he described in detail in *Qu'est-ce que l'urbanisme*? in 1926: "Any city, left to itself, will retain the plan on which it was built. This persistence is only disturbed by local interventions, made known to us by history" (Lavedan 1926a, p. 91). In 1966, the architect Aldo Rossi, speaking of "Poëte and Lavedan's *theory of permanence*", stated: "This last point is Poëte's most important discovery. Cities tend to remain on their axes of development, maintaining the position of their original layout…" (Rossi 1984, p. 59).

On the cusp of the 19th and 20th centuries, the idea that traces of the past survive in the present had thus expanded beyond the point of simply observing ruins and monuments still present in the landscape, to the consideration of broader spatial structures. Several authors used the notion of the "palimpsest" to communicate the idea of temporal collisions, originally in the sense of an accumulation of forms from

---

2 Author's note: for citations from works for which no English edition is available, a translation is provided in the text, with the original in the footnotes. In this case: *…du moins mal connu, qui est généralement aussi le plus récent ou le plus proche de nous, comme par exemple le paysage actuel ou le cadastre du XIXe siècle, pour remonter dans le passé à l'aide d'indices de plus en plus difficiles à interpréter à mesure qu'ils sont anciens, mais dont l'enchaînement historique ne fait pas de doute* .

3 Both authors were founding members of the *Institut d'urbanisme de l'université de Paris* (IUUP), establishing schools of research which were prevalent in typomorphology in both Italy and France in the years following the Second World War (Cohen 1993; Darin 1998).

4 *Un principe, sinon universel et absolu [...], du moins applicable à la majorité des cas : la loi de persistance du plan.*

different periods. This metaphor has proved remarkably durable, used over several decades, with variations in meaning reflecting changes in the way in which we understand the notion of time in a landscape.

### 1.1.3. *The palimpsest as accumulation*

The term "palimpsest" was used from ancient times to denote a tablet or sheet of parchment, which was scraped to remove earlier text prior to reuse. From the Renaissance on, chemical techniques were used to read the undertext of palimpsest manuscripts; these techniques became increasingly sophisticated, reaching their height in the 18th century (Larousse 1898, p. 628; Gaffiot 1981, p. 1105). The metaphor began to be applied to landscape in the 19th century; the first recorded instance is found in the work of F.W. Maitland (Lucas 2012), who studied dispersed habitat of presumably Celtic origins "from the ordnance map (that marvellous palimpsest, which under Dr Meitzen's guidance we are beginning to decipher)" (Maitland 1987, pp. 15–16). Maitland belonged to a school of thought in which landscape was compared to text, and elements in a landscape were analyzed using semiological techniques. In 1934, the historian H.J. Randall again used the palimpsest metaphor, indicating that elements in a landscape should be seen as signs, and that their assembly constitutes a historical document in the same way as written documents. For Randall, maps provided a record of history, inscribed into the landscape:

> The face of the country is the most important historical document that we possess. Upon the map of England – "that marvellous palimpsest" – is written much of English history: written in letters of earth and stone, of bank and ditch, of foliage and crop. As is the case with every map, the writing is not such as he that runs may read. It needs patience to discover, knowledge to decipher, insight, sometimes amounting to genius, to interpret. But the writing is there, all else awaits the competence of the reader. (Randall 1934, p. 5)

For Maitland and Randall, this written past was not concealed or erased by more recent "writings". Ancient forms are present in the current landscape, and can be accessed by anyone who knows how to read them. One must simply look for clues in the image, as in an Épinal print[5], where the most important points are not necessarily the most obvious. The villages which Maitland studied, using maps, were still part of the contemporary landscape, but they had retained elements betraying their ancient origins: groupings of elements, road layouts, housing patterns etc., which "never have been thoroughly effaced" (Maitland 1987, p. 15).

5 Mass-produced prints popular in France in the late 19th century, many featuring a hidden image concealed within the broader view.

Carl Sauer (1889–1975), a contemporary of Maitland, professor of geography at the University of Berkeley and one of the leading figures in the development of historical geography, further disseminated Meitzen's work and promoted the idea of "a picture of the former cultural landscape concealed behind the present one", which could be read by anyone with the right keys (Sauer and Leighly 1963, p. 367).

The idea of the palimpsest is helpful as a metaphor for understanding the coexistence of new and ancient forms in a landscape. The French geographer Paul Vidal de La Blache used the term in the sense of an accumulation:

> Big industry has upset conditions in central and western Europe during the last century. A thousand years of history had made […] erasures on the record of population[6] [original: *Ce peuplement s'offrait déjà comme un palimpseste sur lequel dix siècles d'histoire avaient inscrit bien des ratures*]. Draining of marshes and clearing of forests were continually adding new touches to the original background. […] But when the industrial age began, it gave birth to a whole new set of human establishments. And yet, the primitive core of the population can still be discovered. On positive evidence it can be stated that men, here as elsewhere, persisted in *assembling* in certain places rather than in others. (Vidal de La Blache 1926, pp. 61–62, our italics)

Despite the transformations brought about by industrialization, medieval settlements remain evident on maps, with an accumulation of different forms of habitation on the same ground footprint (Figure 1.2).

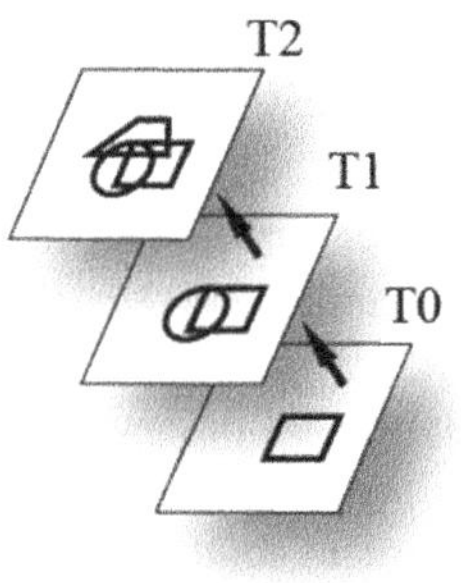


**Figure 1.2.** *Palimpsest as accumulation. The palimpsest came to be used as a metaphor in landscape studies in the late 19th to early 20th centuries, communicating a notion of an accumulation of forms, rather than erasure and replacement (S. Robert 2019)*

6 *Peuplement*: not just "population" in the usual sense of the term, but also "peopling", i.e. settlement, the way in which man inhabits the land.

The palimpsest metaphor has not only been applied to habitat distribution, but also to agrarian structures and to landforms. In 1934, the historian Marc Bloch, using the same notions of reading and deciphering as the English authors cited above, wrote:

> Field layout, perhaps to a greater extent than even the grouping or form of houses, is the book in which rural societies inscribed, line by line, the vicissitudes of their past. Alas, a paleography of this great territorial palimpsest has yet to emerge. Nevertheless, several authors, in recent years, have attempted to decipher a few pages[7]. (Bloch 1934, p. 483)

In the early years of the 20th century, geomorphologists used what they called "palimpsest theory"[8] to describe natural cirques which bear the imprint of successive phases of glacier coverage, even after the glaciers in question have retreated (Jorré 1933, p. 365). In 1934, describing the wind direction recorded in the morphology of sand dunes in the Sahara, the geographer Léon Aufrère spoke of "deciphering the writing on this sandy palimpsest[9]" (Aufrère 1934, p. 130). Jean Demangeot used the palimpsest metaphor in a similar way in relation to the direction of geological folds and the hydrographic network:

> That these ancient, favored lines have survived beyond orogeny, like the hidden letters of a palimpsest, is clear; this is the key fact retained by morphologists, notably in terms of the tyrannical action of these lines on present features. (Demangeot 1943, p. 571)[10]

The use of the word "tyrannical" implies a negative judgment on the part of the geographer; this critical regard with respect to persistence was a salient feature of discussion in the decades following the Second World War (see Chapter 2). During the first half of the 20th century, however, the palimpsest concept was essentially used to understand observations: the past survives in the present, in the ground

7 *Plus encore peut-être que le groupement des maisons ou leur forme, la disposition des champs est le livre où les sociétés rurales ont inscrit, ligne sur ligne, les vicissitudes de leur passé. Malheureusement, ce grand palimpseste des terroirs attend encore sa paléographie. Du moins, de nombreux travaux, au cours de ces dernières années, se sont-ils efforcés d'en déchiffrer quelques pages.*

8 First proposed by the meteorologist Griffith Taylor after 1911, and later introduced in France by Georges Jorré (1899–1956).

9 *…dechiffrer cette écriture sur son palimpseste de sable…*

10 *Ce qui est sûr, et c'est en définitive ce que retiendra le morphologue, c'est la survie étonnante de ces lignes privilégiées anciennes par-delà les orogenèses, telles les lettres cachées d'un palimpseste, et surtout leur action tyrannique sur les traits du présent.*

footprint of certain forms, orientations, directions, etc., even in cases where the initial reasons for this footprint (populations, land administration systems, winds glaciers, etc.) are long departed.

## 1.2. Change, an eternal constant?

### 1.2.1. *The diachronic approach*

The survival of the past in the present was formalized, in scientific terms, in the area of linguistics in the 19th century. In the specific field of historical etymology, the state of a language at a time *t* is considered to be derived from a "source" language, and all linguistic phenomena are considered to "carry within them the trace of their past"[11] (Ducrot and Schaeffer 1995, p. 334). Ferdinand de Saussure refined this definition in relation to time in his *Course of General Linguistics*, introducing the term *diachronie* (diachrony) (de Saussure 1995a, p. 117)[12]. Formed from the Greek *dia*, through, and *chronos*, time, this realm of linguistics describes the successive states of a language and the phenomena which cause a language to shift from one state to another. Above and beyond its methodological implications, diachrony implies a certain vision of time. In Saussure's view, time is responsible for a dual phenomenon of "mutability and immutability", in which signs may be altered by time while retaining certain elements:

> [...] linguistic changes do not correspond to generations of speakers. There is no vertical structure of layers one above the other like drawers in a piece of furniture; people of all ages intermingle and communicate with one another. (de Saussure 1995b, Part I, Chapter II, § 1)

The state of a language is thus, in part, inherited from the language's past. The notion of diachrony can be used to take change into account, while retaining links to the past, using the postulate that certain persistent elements establish a type of continuum across and between different periods.

### 1.2.2. *Connecting past and present*

In common parlance, continuity denotes "unbroken and consistent existence or operation...a connection or line of development with no sharp breaks" (Concise

11 ...*portant en lui la trace de son passé.*

12 Based on lectures given by Saussure, the first edition of the book was published posthumously in 1916.

Oxford English Dictionary, 2008, p. 309). For Saussure, continuity was the first condition for diachrony:

> The sign is subject to change because it continues through time. But what predominates in any change is the survival of earlier material. Infidelity to the past is only relative. That is how it comes about that the principle of change is based upon the principle of continuity. (de Saussure 1995b, Part I, Chapter II, §2)

The survival of initial material means that the sign continues to exist over time, but permits alterations in meaning. Language exists because of the "speaking masses". It is intimately linked to the social mass, which is "naturally inert", and thus acts as a powerful conservatory factor (de Saussure 1995a, p. 112). The idea of permanence is thus linked to the notion of inertia, one of the fundamental principles of 19th-century mechanics, in which it was used to denote "a property of matter by which it continues in its existing state of rest or uniform motion in a straight line [...]" (Concise OED, 2008, p. 727).

Using the notion of continuity, time is seen as duration – *durée* – that is, as something which runs between two defined limits. The French verb *durer*, to last, also implies continuation with the same qualities (Larousse 1922, p. 695). A form of analogy seems to have appeared between the terms *dur*, *dureté* (hard, hardness) and *durée*, duration, from an early date; after all, the words share the same Latin root[13]. The Latin verb *durare*, initially meaning "to harden, to fortify", also took on a temporal sense in ancient times. Lucretius used the word to suggest that the body could not survive for longer than the soul after death. Tacitus used it to report facts that he had been unable to verify personally, but which had been established by authors who themselves were still alive at the time of the historian's youth (Gaffiot 1981, pp. 565–566).

The idea that the permanence of forms is a result of resistance, primarily in the sense of material continuity, is an old one. In 1622, for example, N. Bergier ascribed the continued existence of Roman roads to their construction. Speaking of "Brunehaut" roads, wide Roman thoroughfares found across the former Belgian Gaul, he stated that one hypothesis for the use of the term *via ferrata* (lit. "iron road") in this context is "the hardness and firmness of their construction, which has resisted wear from carts for fifteen or sixteen hundred years"[14] (Bergier 1622, p. 93). In the 19th century, the historian Camille Jullian also emphasized the hard-wearing

13 Author's note: cf. the English term "durable", from the same root, which combines elements of "hardness" with a notion of continuity over time.

14 *...la dureté et fermeté de l'ouvrage, qui depuis quiez ou seize cens ans resiste au froissement du charroy.*

(*résistant*) nature of the materials used in ancient road construction, comparing their foundations to an "invisible wall" (Jullian 1964, pp. 108–109). The continued existence of ancient roads was thus considered to be intimately linked to the notion of physical resistance. The *dureté*, hardness, of their construction ensured their *durée,* perennity, long after Rome had fallen. Material hardness may also be linked to the durability of power. In a proverb quoted by the Venerable Bede in the 7th century, the persistence of Rome, and, by extension, of the whole world, is linked to the resistance of a monument (the Colosseum):

> As long as the Colosseum stands / Rome will stand; But when the Colosseum falls / Rome, too, will fall. And when Rome has fallen / Then the whole world will fall with it.[15] (Du Cange 1678, p. 407, translated)

According to this association of the notions of *durée* and *dureté*, the survival of a form over time results from the continued existence of the initial material object, if only in the form of ruins or traces. This assimilation of physical solidity with the capacity to last over time (durability), or even with the capacity to ensure the perennity of a societal choice, is at the heart of current debates on the subject of resilience; the approach is at odds with the notion of engineering resilience, where the response to natural disasters is found in the physical resistance of infrastructures (section 5.2.4).

In the early 20th century, certain authors observed that the survival of forms in a landscape also relies on adjustments, whereby the original material forms are adapted to new functions. For example, it was noted that the ground footprint of medieval ramparts, later demolished, could still be seen in the curvature of urban thoroughfares (Brunhes 1925; Lavedan 1926a; Unwin 1981; Rossi 2001). P. Lavedan, writing during the period of reconstruction which followed the First World War, noted that towns or cities destroyed by a violent event (fire, war, etc.) grew back "naturally" following their earlier layout. Lavedan considered this phenomenon to be the result of a "natural inclination", whereby "an owner will spontaneously rebuild his dwelling where it was before". Once again, continuity appears as the result of memory and of uninterrupted ownership. Furthermore, in cases where destruction and reconstruction are separated by a longer period, "experience shows that [...] in many cases, if even traces of a house remained, it will be rebuilt in the same place"[16] (Lavedan 1926a, pp. 92–93). Material continuity

15 *Quandiu stabit Colyseus – stabit et Roma; quando cadet Colyseus – cadet et Roma; quando cadet Roma– cadet et mundus.*

16 *...sentiment naturel... [...] le geste spontané du propriétaire est de rebâtir sa demeure où elle était. [...] L'expérience montre [...] que dans beaucoup de cas les maisons, pour peu qu'il en ait subsisté quelques restes, sont ensuite relevées à la même place.*

thus takes the place of memory. For Lavedan, permanence was the rule, while the creation of something new was the exception. The "natural inclination" of the owning masses, spontaneously rebuilding their homes in the same locations, thus corresponds to Saussure's "speaking masses".

The argument linking the permanence of ownership with a form of fixation of plot layouts through construction echoes the connection between *durée* and *dureté* described earlier. In the 1970s, the architect Pierre Pinon described the way in which certain Roman amphitheaters were progressively replaced by dwellings built on top of the ancient foundations (Pinon 1978). The parcel system constitutes a record of ways in which the land was used, even after it has ceased to accomplish a given function. Plots illustrate the contours of a monument, and these lines are "hardened" by buildings on these foundations. The footprint of extinct material structures thus "grows up" from the ground (Pinon 1994, p. 40)[17]. Describing the "mechanisms of memory in the parcel system", the architect Bernard Gauthiez highlighted the links between legal and material criteria. He assigned a major role to the influence exerted by public power, guaranteeing the persistence of public land boundaries, while also noting that the persistence of boundaries is "strengthened by the tendency in urban settlements, at least from the late Middle Ages on, to construct dwellings along boundary lines"[18] (Gauthiez 1999, p. 19).

The persistence of older forms is not, therefore, exclusively a result of the persistence of the constructions themselves (shapes). Nevertheless, up until the 1990s, researchers still struggled to explain this persistence without resorting to the notion of continuum. There is an uninterrupted link between past and present, the result of multi-generational memory or, in the longer term, of the persistence of material remains, functions and the cadastral plan (Robert 2003a). This approach excludes the possibility of temporal hiatuses or non-linear temporalities.

### 1.2.3. *Drivers of change*

Nevertheless, these early morphologists were well aware of the existence of change. This is evident in reactions to Meitzen's suggestion that 19th-century parcel arrangements were a direct reflection of divisions established centuries earlier. Marc Bloch drew inspiration from Meitzen's work, but situated his regressive method within a more dynamic view of history. Refusing to see the past as a still photograph, Bloch preferred a cinematic metaphor:

---

17 For the analysis of the permanence of Roman amphitheaters in urban plots, see also Lefebvre (2001).

18 *...d'autant plus efficace qu'en milieu urbain, du moins à partir de la fin du Moyen Âge, les maisons sont souvent construites en limite séparative.*

> If we use our common sense, we shall see that the picture presented by the recent past is not an image we merely need to project over and over again in order to reproduce that of centuries more and more remote; what the recent past offers resembles rather the last reel of a film which we must try to unroll, resigned to the gaps we shall certainly discover, resolved to pay due regard to its sensitivity as a register of change. (Bloch 1966, p. xxx)[19]

If, then, the landscape is in a state of perpetual change, how can we justify the use of cadastral plans, modern or contemporary maps to study ancient or medieval forms? According to A. Verhulst, these documents constitute a valid source in cases where there is no doubt in the matter of historical sequentiality (Verhulst 1995, p. 20). In order to deduct the past from the present or, conversely, the present from the past, the observed state must be precisely situated within a logical progression. The feasibility of regressive studies is thus contingent on the conceptualization of this progression or sequence.

In the 19th and the first half of the 20th century, evolutionism, as defined in the natural sciences, was seen as a coherent framework for describing change. Social scientists used these tools to compare different types of civilizations, based on an idea of progress, defined by Enlightenment philosophers as the ultimate goal for humanity. Evolution appeared to offer the means of attaining this goal, by means of a collective movement through a chain of states, with each clearly situated on a linear and cumulative timeline. Given a logical organization of time into a series of periods, it would then be possible to work backwards. In the words of Marc Bloch:

> Let us then agree, since we have no choice, to follow the trail backwards, one careful step at a time, examining irregularities and variations as they come, avoiding the all-too-common error of trying to leap at a bound from the 18th century to the Neolithic age. (Bloch 1966, p. xxx)[20]

---

19 *Au proche passé, la méthode régressive, sainement pratiquée, ne demande pas une photographie qu'il suffirait ensuite de projeter, toujours pareille à elle-même, pour obtenir l'image figée d'âges de plus en plus lointains ; ce qu'elle prétend saisir, c'est la dernière pellicule d'un film, qu'elle s'efforcera ensuite de dérouler à reculons, résignée à y découvrir plus d'un trou, mais décidée à en respecter la mobilité.*

20 *Suivons, puisqu'il le faut, en sens inverse la ligne des temps ; mais que ce soit d'étape en étape, attentifs toujours à tâter du doigt les irrégularités et les variations de la courbe et sans vouloir – comme on l'a fait trop souvent – passer, d'un bond, du* XVIIIe *siècle à la pierre polie.*

Bloch's reference to *pierre polie* in the original French (translated here as "Neolithic") is significant, highlighting the role of technology as an essential factor in the attainment of different developmental stages (Latour 1991). This is further evident in the 19th-century classification of ancient societies: "stone age", "iron age", etc. Material and technical characteristics were essential criteria, used in the emerging sciences of ethnology and archeology to define different cultures.

The notion of age is ever-present in the study of morphology; some researchers expressed concern that the continued survival of past forms into the present might hinder the attainment of an optimum state by means of evolution. For Marcel Poëte, there was:

> [...] a difficulty in reconciling the highly-evolved living conditions of the modern day with these persistent remains. How can ancient country paths, evolving into roads, be suitable for automobile circulation, in a city which has, itself, lasted from a past agricultural age to the current age of powerful industrialization without casting off its old skin? (Poëte 2000, p. 23)[21]

The notion of age is also integral to the cycle of erosion concept in geomorphology, introduced by William M. Davis (1850–1934) and popularized in France by Henri Baulig (Masutti 2002). In this idealized model, landforms develop through a series of evolutions, step by step, finally resulting – when conditions allow – in a peneplain. The erosion cycle concept provides a means of explaining evolutions in a landscape. Under the effects of erosion, relief goes through a number of different "ages", from youth to maturity and onto old age. Davis believed that, using this model, the past and future states of a relief form could be predicted based on observations of a present state (Meynier 1969, p. 58).

At the same time, the image of the living organism was also being used to describe towns or cities, seen to traverse a series of ages. Poëte, for example, introduced an idea similar to those used in geomorphology, the "vital cycle"[22], in the context of urban development:

> In studying this city in the past, we do not look at a skeleton, but rather at a living being, younger in age than at present. The city, as a

21 *...la difficulté qu'il y a à concilier les conditions de la vie très évoluée de nos jours avec ces persistantes survivances. Comment d'anciens chemins ruraux, mués en rues, peuvent-ils se prêter à la circulation automobile, dans une cité qui, de l'âge agricole d'antan, est passée, sans faire peau neuve, à l'âge puissamment industriel d'aujourd'hui ?*

22 In reference to Henri Bergson's notion of *élan vital*, which had a considerable influence on Poëte's work (Guillot 2014).

> collectivity, is subject to the effects of time over the course of its existence, in the same way as an individual; at any given time, it has an age. The notion of the vital cycle, used in geography with respect to the landforms, is thus crucial to the study of urban settlements. (Poëte 1924, p. 1)[23]

Using this approach, Poëte analyzed the dynamics of a town or city by comparing it to a living being. Elements are transformed in one direction, from birth to death, making it possible to "discern the degree of evolution" at any given time (Poëte 2000, pp. 83–84). This logical sequence of known steps (birth, maturity, death) can be followed in both directions, to travel back in time (regressive history) and to improve our understanding of the present.

Thus, historians may use the present to understand the past in the absence of source material for older periods. For geographers, on the other hand, incursions into the past give a better understanding of the present, and only the elements which are helpful for this purpose will be retained. In history, this approach is called "regressive", whereas in geography, the term "retrospective" is used (Dion 1949). In both cases, time is considered as a continuous line, in which the present never completely eliminates legacies from the past. Past and present are linked by a logical chain, as Jean Brunhes and Camille Vallaux put it in their 1921 work examining the relationship between geography and history. For these authors, history offers the means of placing an isolated fact "into the stream of life which produced it", as "one link in a chain" (Brunhes and Vallaux 1921, p. 21).

### 1.2.4. *Primary forms: a non-evolving landscape*

From the outset, however, the regressive analysis approach to history was confronted with a difficulty: that of studying landscapes which are in a constant state of evolution. Without a general framework for studying the evolution of human societies, it is hard to interpret plans or documents beyond their date of production. This led to a focus on planned parcel systems which marked a clear change from previous landscape patterns, for example following periods of land clearance. From a methodological perspective, a distinction was made between two types of landscape, of which A. Verhulst summarized the characteristics.

---

23 *L'étude de cette cité dans le passé n'a point pour objet un squelette, mais un être vivant qu'on prend seulement à un âge antérieur à celui qu'il a présentement, car la collectivité qu'est la ville subit, au cours de son existence, tout comme l'individu, l'action du temps et a, au moment où on la considère, un âge donné. Il importe d'introduire, dans l'étude de l'agglomération urbaine, la notion de cycle vital, de même qu'en géographie on l'a introduite dans l'étude par exemple des formes du relief du sol.*

A landscape is called "primary":

> ... when it has undergone no major changes since its formation at a given moment in the Middle Ages [...] In this case, as a general rule, it will have been formed over a relatively short period of time and following a pre-established plan, most often in a context of collective clearing. (Verhulst 1995, pp. 48–49)[24]

The degree to which land systems were planned appears to have played a role in their fixation, limiting possibilities for future evolution, notably in the direction of plot inequality, since the plan covered the entirety of the townlands before they were put into agricultural use. Furthermore, the length and breadth of parcels were pre-defined from the outset. The limited scope for evolution in such cases facilitates reconstitution on the basis of later sources (Verhulst 1995, pp. 48, 52–53). In this case, the hardness (*dureté*) indicated by the persistence of architectural constructions (section 1.2.3) is a result of the planning process. As the historian Raymond Chevallier wrote in 1958:

> In reality, centuriation is like all of the monuments which come down to us from ancient times: it passes through time unscathed; this is both its strength and its weakness, and this permanent character should be seen as a hallmark of Rome. Centuriation, born of the earth, was not a stone corset preventing it from breathing, but it did assign it a structure. The system gave it a stability, making it immune to successive recombinations and dismemberments. It only disappeared once it was no longer maintained, fragmenting much faster in regions artificially clawed back from the desert, where man's continued survival is only attained through constant struggle.[25] (Chevallier 1958, p. 121)

---

24 *...lorsqu'il n'a pas subi de grands changements depuis sa formation à un moment donné du Moyen Âge [...] sa formation, dans ce cas, aura eu lieu, en règle générale, pendant un laps de temps assez court et selon un plan préconçu, le plus souvent dans le cadre d'un défrichement collectif.*

25 *En réalité, la centuriation est comme tous les monuments que l'Antiquité nous a laissés : elle brave le temps qu'elle ignore ; c'est là sa force et sa faiblesse, et c'est dans ce caractère de permanence qu'il faut voir la marque propre de Rome. La centuriation, née de la terre, n'était pas un corset de pierre qui l'empêchait de respirer, mais elle lui conférait une structure. Elle lui a donné une stabilité qui lui a évité les remembrements et les démembrements successifs. Elle n'a disparu que lorsqu'on a cessé de l'entretenir et son effritement a été d'autant plus rapide qu'il s'agissait de régions artificiellement conquises sur le désert toujours menaçant, où l'homme ne peut se maintenir que par une lutte constante.*

In contrast, a "secondary" landscape is one which has undergone:

> ...numerous transformations [which] occurred over a period, often several centuries long, which changed the primary form to an extent which renders reconstitution necessary, laborious, difficult and, in most cases, only partially possible...[26] (Verhulst 1995, p. 49)

For methodological reasons, the study of secondary landscapes has been largely abandoned due to its complexity, in favor of studying planned parcel systems: medieval urban foundations, parcel systems resulting from land clearances, rural parcel systems in Eastern France, etc. These specific town and land plans are still "readable" in the present, having undergone very limited degradation due to their original coherency. In this approach, time is seen as a framework which is external to the observed object, and acts on elements in the sense of alteration.

### 1.2.5. *The notion of decay*

In the 19th century, linguists developed a theory of linguistic decline, whereby all languages are considered to be derived from a proto-human language (Ducrot and Todorov 1972). For Saussure, time is a universal law, which acts on language just as it acts on any other element: "For time changes everything. There is no reason why languages should be exempt from this universal law... This evolution is fatal" (de Saussure 1995b, Part I, Chapter II, § 2).

A key goal in landscape studies is to identify the initial finished state of forms (a kind of "mother form"). All later additions or transformations are thus considered as disturbances or degradations to the original plan. For example, Bloch highlighted the case of drawings of fields or monuments which only very rarely survive to the present day in their "pure" form (Bloch 1988, p. 51). Maitland used the terms *true village* and *purest form* to describe a grouped village settlement, established at the time of the Germanic Conquests in the early Middle Ages (Maitland 1987, pp. 15–16). In this approach, there is an initial, finished, "pure" form which is then altered by time over the centuries. In ecology, the notion of a "pristine" ecosystem, untouched by human hands, plays a very similar role (see Part 2). This primal form, which, as we shall see, is more of an ideal type than an element which actually existed, is clearly situated in the past: in the present, we can only access a decayed image. The regressive approach aims to work backwards through time in order to reconstruct a "least decayed" version.

---

26 *...de nombreuses transformations [qui] sont intervenues pendant une période parfois longue de plusieurs siècles et qui ont changé la forme primaire dans une mesure qui rend la reconstitution nécessaire, laborieuse, difficile et le plus souvent même partielle...*

A major aim in morphology, for many years, was to conceptualize modes of decay rather than those of the persistence of past forms. For example, R. Chevallier proposed the criterion of "visible wear" for dating visible objects in a landscape:

> Using the criterion of "visible wear", within the framework of a morphological series, the structure which presents the most worn appearance is likely to be the oldest. Objects in a series may be arranged in order of decreasing sharpness: structures can be more or less "fresh", more or less decayed by the effects of time, to the point where they become like ghosts… (Chevallier 1971, p. 108)[27]

In 1983, Gérard Chouquer proposed a model for explaining the degradation and fossilization of centuriated cadastral plans (Chouquer 1983). According to this model, the original, regular centuriations were transformed as a result of the polarization of the road network around medieval grouped habitats (Figure 1.3)[28].

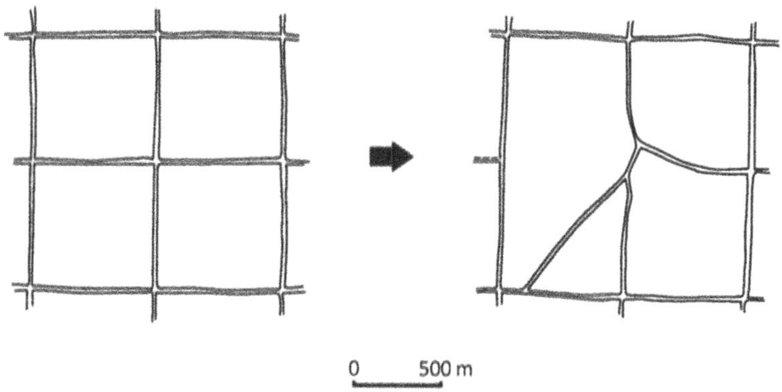


**Figure 1.3.** *Theoretical illustration of the decay of a centuriated cadastral system (inspired by Chouquer 1983; Robert, 2020)*

Until the 1980s–1990s, morphologists aimed to identify the traces of plans and parcel systems, based on a preconceived notion of the finished form which they were presumed to have in ancient and medieval times. This mother form was used as

27 *En vertu du critère "d'usure apparente" utilisable dans les séries morphologiques, la structure qui présente l'apparence la plus usée risque d'être la plus ancienne. Dans une même série, on peut ranger les objets par ordre de netteté décroissante : les structures pouvant être plus moins fraîches, plus ou moins dégradées dans le temps et rendues comme fantomatiques…*

28 In the 2000s, the author developed a contrasting idea that medieval and modern habitats play a key role in the maintenance of ancient forms (section 3.2).

a reference for hypothetical reconstructions of missing sections on maps. In the context of the ancient world, many methods used standardized grids based on the most common dimensions of contemporary planned parcel systems, transferred onto an overlay then held up against topographical maps or aerial photographs (Chouquer 1990). For ancient urban environments, P. Pinon pioneered the use of "programmatic plans" using a unique, module-based orthogonal grid, based on current city layouts (Pinon 1994)[29]. This approach consists of looking for traces on modern maps and photographs that correspond to the grid layout. However, it leaves little room for the geographical particularities of specific sites, which may warp the grid; the grid takes priority over the actual forms of the site. Studying planned layouts in this way implies a minimization of the effects of both time and space.

## 1.3. Reversible time

### 1.3.1. *Resistance to change*

During the 19th century, the development of research into planned layouts in the historical sciences ran parallel to a growing interest in ancient plans and morphological analysis within the emerging field of urban design. This interest appears to have drawn on a critical approach to industrial-era town design, in which settlements expanded beyond historical limits in the form of medieval and modern walls (Choay 1965; Cohen 1993). The desire to act on the forms of towns and cities went hand-in-hand with a desire to maintain a "legacy" state, or even to recreate an initial state from a contemporary "decayed" state. Certain town planners believed that the application of a historical plan would rejuvenate, or even resurrect, a former state. In 1936, for example, P. Lavedan criticized the vitalism present in his own earlier words as it did not correspond to planned cities:

> [...] historical fatalism has its roots in the assimilation of the city to a living being. A city is seen as a living thing which, like all living things, is born, grows, and dies; it is a child, an adolescent, then an old man. I myself accepted this comparison for many years, even in the first edition of this work; now, however, I find it to be unacceptable. [...] On the contrary, a city can be rejuvenated or even resurrected. While all men start out as children, many towns do not have a "childhood", being in possession of their full strength from the outset, like Athena springing fully-formed from the mind of Zeus; the

29 For a critical assessment of these methods, see Chouquer (1994) and Favory (1997).

> existence, or even the very notion, of these planned cities has no place in the fatalist approach. (Lavedan 1959, p. 13)[30]

Among urban planners, there was an idea that certain types of town plans provided a means of combatting aging and decay. Michel Parent, museum conservator of relief plans, writing in 1948, considered that the knowledge of ancient could be seen as an "applied science of curation":

> The past can serve the future in all cases, on the condition that we do not make a tyrant of it: firstly, it highlights tried-and-tested postulates or permanent laws which are helpful in finding solutions to new problems; secondly, in specific cases, awareness of a pathological evolution of a city or district enables us to distinguish between healthy and sick parts, providing urban planners with genuine remedies.[31] (Parent 1948, p. 282)

These approaches place a particular emphasis on one time of plan: that of planned cities, considered in this case of works of art.

### 1.3.2. *The ancient plan as a work of art*

Toward the end of the 19th century, the Austrian architect Camillo Sitte, in *Der Städtebau* – published in English as *City Planning According to Artistic Principles* – emphasized the esthetic qualities of the medieval town, "rediscovered" by the Romantic movement (Sitte 1890). In Germany, Sitte's contemporaries, the architect Joseph Stübben and the art historian Albert Erich Brinckmann, highlighted the

30 *[...] le fatalisme historique a pour base l'assimilation de la ville à un être vivant. La ville, dit-on, est un être vivant et, comme tous les êtres vivants, elle naît, elle grandit, elle meurt ; elle est tour à tour enfant, adolescent, vieillard. La comparaison, que j'ai longtemps acceptée moi-même, ne serait-ce que dans la première édition de cet ouvrage, me paraît aujourd'hui inacceptable. [...] La ville, au contraire, peut rajeunir ou même ressusciter. Surtout, alors que l'homme commence par être enfant, beaucoup de villes n'ont pas eu d'enfance, sont apparues du premier coup en pleine force, en possession de tous leurs moyens, comme Athéna jaillit toute armée du cerveau de Zeus : ce sont les villes créées, dont les théories fatalistes ignorent l'existence et nient même implicitement le principe.*

31 *Le passé peut servir l'avenir de toutes les manières, à condition qu'on n'en fasse pas un tyran : d'une part il permet de dégager des postulats ou des lois permanentes qui ont fait leurs preuves et qui aident à trouver les solutions des nouveaux problèmes ; d'autre part, dans les cas précis, la connaissance de l'évolution pathologique d'une ville ou d'un quartier aide à distinguer le membre sain et le membre malade et met dans les mains de l'urbaniste les vrais remèdes.*

interest of studying ancient city plans in order to identify models of successful urban arrangements (Stübben 1890; Brinckmann 1908). In England, Raymond Unwin demonstrated that ancient urban forms provided a valuable source of models for town development (Unwin 1981). In 1926, Lavedan aimed to lay scientific foundations for this approach in his seminal essay *Qu'est-ce que l'urbanisme*. The publication was accompanied by a selection of ancient, medieval and modern plans; the way in which certain forms, such as the checkerboard, appeared again and again was seen to be particularly notable (Lavedan 1926a, 1926b, 1941). This approach, labeled "culturalist" by the architectural historian Françoise Choay (1965), emphasized ancient city layouts with a certain esthetic value, giving them the potential to be used as models in developing an art of urban design.

In order to establish his history of urban architecture as a science, Lavedan needed to define a position for it among all of the other disciplines for which the town was an object of interest in the early 20th century (geography, sociology, economics, history, etc.). He chose to focus on one aspect of urban studies which, in his view, had been largely ignored in other disciplines: the notion of the plan from the perspective of material elements (road traces, arrangement and specialization of different districts, design of public squares and parks). However, given the proximity of this approach to geography, Lavedan chose to focus only on urban plans with a certain esthetic value, considered from an artistic perspective, in order to "establish, with respect to related disciplines, the possibility of an independent discipline in the history of urban architecture, written from a purely artistic and esthetic perspective" (Lavedan 1926a, p. 7)[32]. It was plans of this type that Lavedan selected for publication in later collections (Lavedan 1926b, 1941).

### 1.3.3. *The ancient plan as a societal project*

At its heart, 19th and 20th century urbanism carried the idea that the choice of a type of plan or building scheme could promote the development of a societal model. The urban plan was seen to have the potential to establish a spatial order that would also imply a social order. Notions of architectural disorder are often closely linked to ideas of social disorder in discussions of urban design (Poïesis 1998). Up until the 1970s, urban design projects were still being presented as "harbingers of spatial order, opposing urban chaos and fragmentation" (Choay and Merlin 1988, p. 625). For the culturalists, the idea of reviving a former, "ordered" state was closely linked to a desire to normalize "chaotic" industrial cities (the same type of logic is still applied in the context of discussions surrounding suburban development, city outskirts, etc.). The esthetic qualities inherent in certain older configurations had

32 *...marquer, vis-à-vis des disciplines connexes, l'indépendance possible d'une histoire de l'architecture urbaine, écrite dans un esprit purement artistique et esthétique...*

been confirmed over the course of history, an opinion supported by the recurrence of these arrangements, confirming the intrinsic value of the plan. Thus, the mother form was also considered as a social project. According to this perspective, space is seen as a relatively neutral substrate, a canvas for the models applied to them by urban designers – models that could be adapted to any location and time, with no need to account for the specific characteristics of geographical sites or for the unique trajectory of a place. On the contrary, dissociation from the natural milieu was seen as a potential criterion for ensuring the durability of an urban form.

### 1.3.4. *Artificial versus natural forms*

From an early date, observations of persistence were linked to the idea that manmade forms had a greater capacity to withstand the test of time than natural forms, simply due to the role played by human rationality in their creation. For N. Bergier, ancient roads survived primarily due to the role of human reason:

> The form of the great highways is artificial, consisting of an assembly and arrangement of the aforementioned materials in a certain order, invented by human industry through the use of reason: not only to create them, but also to preserve them as long as the craftsmen's art and the nature of the materials themselves would permit.[33] (Bergier 1622, p. 135)

Bergier's whole work was devoted to the development of an ideal model of the ancient road, which he presented as a veritable architectural order (Robert and Verdier 2014, pp. 11–71). He created a classification of highway-building materials using a scale of values, from the least useful natural deposits to the most sophisticated man-made materials, such as "Tiles [...] formed not by the hazards of time, but using rule and compass"[34] (Bergier 1622, p. 193).

The same type of opposition resurfaced three centuries later in the work of Lavedan, who suggested that towns could be split into two categories: "spontaneous towns, born of chance and which grew up gradually, and artificial towns, created in a single day by the will of one man"[35]. In the first case, he considered that towns

33 *La Forme des grands Chemins est artificielle, consistant en l'assemblage et disposition des Matieres que dessus en certain ordre, que l'industrie humaine a inventé par raison : non seulement pour donner l'estre à l'ouvrage : mais aussi pour le conserver plus longtemps que l'art des Ouvriers et la Nature des materiaux le peuvent porter.*

34 *Quarreaux [...] qui ne sont formez à l'advanture : mais à la regle et au compas.*

35 *...les villes spontanées, nées du hasard et qui ont grandi peu à peu, les villes artificielles, créées en un jour par la volonté d'un homme.*

"left to chance, or to nature, the task of grouping the component elements around the generating element" (road, watercourse, relief, etc.), while artificial towns were "constructed following a predetermined plan". He added that the latter represent "the specific object of study of a history of urban architecture: not always works of beauty, but always works of art, or in other terms, intentional creations of human ingenuity"[36] (Lavedan 1926a, p. 5). Lavedan established a hierarchy of plans based on their independence with respect to the natural dispositions of the host site and on their geometric complexity: from the "inorganic village, in which dwellings seem to have sprung up at random" to "the checkerboard plan, with a grid pattern which is now the boast of many cities of the New World, from Buenos Aires to Chicago"[37] (Lavedan 1926a, p. 31). The artificial form as a work of art, particularly the regular geometric plan, was seen as a product of human rationality forging a structure with the capacity to slow the decay of forms under the influences of time and nature (section 1.2.5). Thus, while M. Poëte emphasized the role of natural sites and of the location of towns within a network of transport arteries in the context of urban planning, P. Lavedan developed an esthetic vision independent of geographic realities. This approach left an indelible mark on the study of urban morphology, both in architecture and archeology.

### 1.3.5. *Conclusion*

In the 19th century and the first half of the 20th century, the past was seen as forming an integral part of the present in landscape. Diachrony and the palimpsest metaphor offered two means of understanding the coexistence of different temporalities based on the notion of accumulation. The diachronic vision, however, revolved around the idea of an initial, finished form subject to decay over time (Figure 1.4). From a scientific perspective, initial states are reconstituted using the regressive method, removing later elements that blur the view. In the absence of a clear logical framework for the evolution of landscape forms, however, this method is difficult to use except in identifying forms which, in theory, have undergone few transformations. Morphologists thus focus on the study of planned forms, which seem the most likely to be well-preserved, due to their independence from the geographical substrate and their structural "hardness" (*dureté*). The culturalist school chose to focus on these forms due to their esthetic properties, and because

36 *...abandonné au sort ou à la nature le soin de grouper les éléments qui les constituent autour de l'élément générateur...construites suivant un plan arrêté d'avance...l'objet propre d'une histoire de l'architecture urbaine, œuvres de beauté, pas toujours, mais œuvres de l'art, c'est-à-dire du génie humain conscient de son dessein.*

37 *...village inorganique, dans lequel les habitations semblent s'être élevées au hasard... plan en échiquier, au quadrillage dont s'enorgueillissent aujourd'hui maintes cités du Nouveau Monde, Buenos Aires ou Chicago.*

they appeared most likely to guarantee, or even restore, a certain social order. As we shall see, this notion, with its inherent idea of temporal reversibility, is not far distant from that of engineering resilience: an important element in current debates on the use of resilience (section 5.2.4).

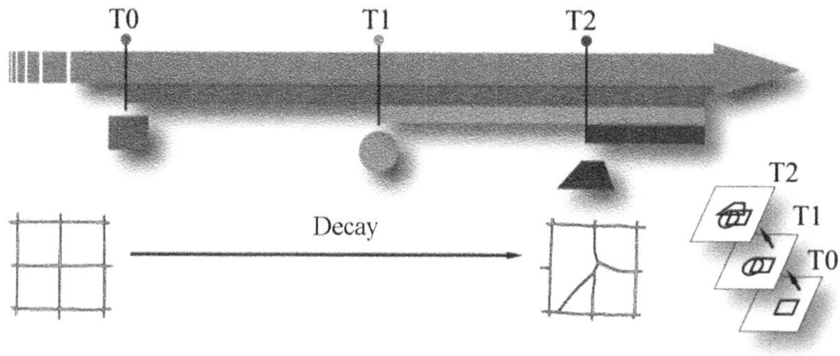


**Figure 1.4.** *Diachrony in landscape forms. The present landscape is seen as an accumulation over time. The past, complete initial form is visible in the present, but in a decayed state. It can be reconstructed by removing later forms (regressive method). The point of reference forms part of a "resistant" past, to which it is theoretically possible to return (Robert, 2020)*

# 2

# Landscape: A Past... Surpassed?

In the years following the Second World War, a new vision came to dominate approaches to landscape: that of a landscape no longer in accordance with the needs of society, even to the point of hindering desires for progress. Where previously the past had been seen as a source of models to imitate, or as the point of origin of ideal forms transmitted to the present and which should be preserved from slow decay, it was now perceived as a factor of inertia, an element to overcome in order to bring about a complete transformation of society. Gradually, the very perception of temporalities in a landscape altered. The past had ceased to be "present", but instead was believed to be buried under the modern landscape; specific techniques would be required to bring the past back to light. The emergence of this view coincided with the development of aerial photography, which offered a new means of observing landscape. From the First World War on, this technique was considered essential in the domain of cartography; geographers and archeologists began to make use of aerial photography during the 1920s. By the 1960s–1980s, the medium had a decisive influence on perceptions of change in landscape.

## 2.1. A visual revolution

In 1955, in the introduction to their *Atlas aérien de la France*, geographers Mariel Brunhes-Delamarre and Pierre Deffontaines highlighted the enthusiasm of researchers for "new views" of the Earth from above, calling the development a "visual revolution":

> The airplane has brought about a complete change in the way we view the Earth, a change greater still than that experienced by the first navigators when they first saw land from the sea. With the airplane, the linear, ground-level view has been replaced by an overview of a whole surface, or even whole volumes; it gives us a new perspective on the

Earth, and, in all truth, is the most extraordinary source of knowledge, to the point where the earth, seen from an airplane, seems like a new planet[1] (Brunhes-Delamarre and Deffontaines 1955, p. 7).

### 2.1.1. *A new vision of landscape*

The use of aviation for military purposes increased rapidly during the First World War, and with it that of aerial photography, used to observe and locate enemy positions. In the early years of the 20th century, many European states launched major national cartography initiatives, producing large scale of their territory using photogrammetric techniques (Lassalle 1990). Surveying was carried out in a systematic manner, using overlapping bands to cover a continuous space, generally equivalent to one panel in the existing topographic map. Terrain was photographed vertical to the camera lens, resulting in a geometric vision of the terrestrial surface, that is, maintaining a constant relation between the photograph and the land. Vertical aerial photography can thus be used as a support for a scientific morphological approach, while aerial photographs taken from an oblique angle cannot be used in this way and only capture observations of the terrain at a given moment in time.

Early in the 20th century, geographers were already well aware of the potential offered by this new tool in studying landscape (Besse 2009, p. 85). Marie-Claire Robic notes the presence of aerial photographs in works by geographers such as P. Vidal de La Blache, Emmanuel de Martonne, J. Brunhes, André Cholley, etc., as early as 1908; a similar interest in aerial views is evident in French, American and British geographical journals from the period 1919–1921 (Robic 2012, pp. 131–136). In archeology, photographs taken from tethered balloons were already being used in the 1870s, and here, too, the new technique was rapidly adopted. Many of the earliest aerial photographs used in this context in the first years of the 20th century were of vestiges of ancient civilizations found in deserted areas of the Mediterranean basin. Theodore Wiegand, director of the Germano-Turkish archeological commission, for example, carried out prospection activities in the Sinai region in 1918 (Chombart de Lauwe 1948, p. 255). In 1919, the British lieutenant-colonel G. E. Beazeley published an article on aerial photography in Mesopotamia in the *Geographical Journal* (Beazeley 1919). Antoine Poidebard, a French priest and air force reservist, carried out a campaign of surveys between 1929 and 1932, resulting

1 *L'avion a réalisé un changement total dans la manière de voir la terre, un changement plus grand que celui que les premiers navigateurs saisirent lorsqu'ils virent la terre pour la première fois de la mer. L'avion a remplacé la vision linéaire et à ras du sol par une vision en surface, ou même en volume ; il a apporté un nouveau point de vue sur la terre et il représente, au vrai, le plus merveilleux mode de connaissance, au point que la terre apparaît, vue d'avion, comme une nouvelle planète.*

in the publication of *La Trace de Rome dans le désert de Syrie*. From 1932 to 1939, Poidebard continued his activities with the support of the French aviation ministry, aiming at perfecting aerial photography techniques for desert steppe areas; his findings in Syria were published in 1945 under the title *Le Limes de Chalcis* (Poidebard 1934; Poidebard and Mouterde 1945).

These earliest efforts in aerial archeology focused on identifying sites by sight, by means of photographs taken at low altitude and from an oblique angle. Sites partially covered by sediments following desertification of the surrounding area, but still visible from above in the form of ruins or microreliefs, were a particular focus. R. Mouterde and A. Poidebard described the citadel site at Zebed, photographed in 1938, as "a field of ruins, hidden in a fold in the terrain" (Figure 2.1). The ring walls "appear as a bulge in the terrain" (Poidebard and Mouterde 1945, pp. 161–162). Aerial vision makes it easier to recognize and "read" the organization of vestiges; viewing a site from above, it becomes possible to understand arrangements which, at ground level, do not present any evident coherence.

**Figure 2.1.** *Aerial photograph of the citadel at Zebed in 1938 by Antoine Poidebard (source: Poidebard and Mouterde 1945, plate XG)*

In Europe, the development of aerial prospection is intrinsically linked to developments in cartography, and the choices made by map-markers concerning the indication of archeological elements provide an insight into the way in which they understood temporalities in the landscape. In the United Kingdom, the Ordnance Survey included archeological data on topographical maps from the 18th century on. For example, the first edition of the One-Inch sheet published on August 14, 1817, indicates the presence of several protohistoric burial sites north-east of Salisbury, identified by engineers during the cartographic survey. This tradition was still alive and well in the period following the Second World War (Figure 2.2).

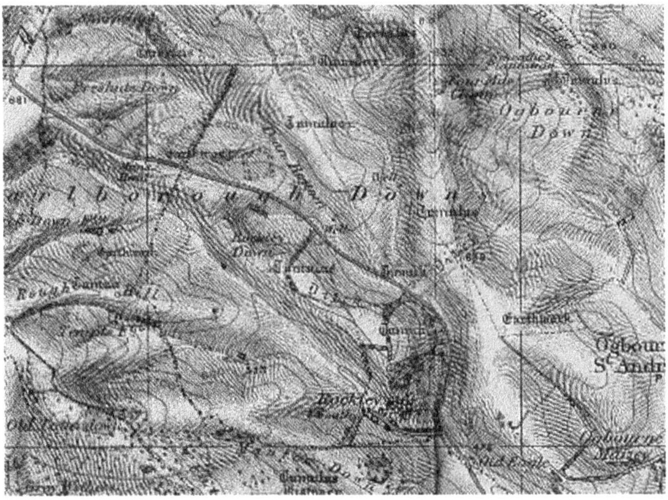

**Figure 2.2.** *Indications of prehistoric archeological features (earthworks, tumuli) on the One-Inch Ordnance Survey map, Marlborough sheet, 1919. These remains are shown in dotted lines and in a different font (source: wikimedia.org/wikipedia/ commons). For a color version of this figure, see www.iste.co.uk/robert/resilience.zip*

In 1920, the Ordnance Survey even appointed an Archeology Officer. The post was first held by Osbert Guy Stanhope Crawford (1886–1957), who was responsible for verifying the precision of archeological indications on the maps released to the general public, to carry out revisions where needed for new editions, and, if possible, to make new discoveries. An Oxford geography graduate, O.G.S. Crawford already had archeological experience prior to his recruitment as a maps officer in the Third Army, then as an air observer with the Royal Flying Corps (Crawford and Keiller 1928, p. V). Crawford is traditionally considered the founder of aerial archeology, of which he developed the theoretical and methodological foundations (Randall 1934, p. 8; Guttmann-Bond and Kluiving 2015, p. 16). At the Ordnance Survey, O.G.S. Crawford launched a series of aerial archeological surveys, the results of which were published from 1923 onward in geographical journals, and in the journal *Antiquity* which Crawford himself founded in 1927 and which became a point of reference in archeology. During his flights, he identified traces likely to correspond to earlier field boundaries; these he presented, under the label "Celtic fields", at the Royal Geographical Society in 1923. The plan of the traces was published in the *Geographical Journal* (Crawford 1923), overlaid on a

1:20,000 map, demonstrating the coexistence of contemporary and so-called "Celtic" fields, tumuli, roman roads etc., shown in different colors (Figure 2.3).

**Figure 2.3.** *Plan of the traces identified by O.G.S. Crawford in the course of his aerial surveys, shown on a 1:20,000 map and described as a "Celtic fields system in Central Hampshire as revealed by air-photographies", published in the Geographical Journal (source: Crawford 1923). For a color version of this figure, see www.iste.co.uk/robert/resilience.zip*

The contribution made by O.G.S. Crawford and other aerial surveyors active in northern Europe in the first half of the 20th century, such as Allen (1891–1940), J.K. St. Joseph (1912–1994), J. Bradford (1918–1975) and J.A. Brongers (1933-), was a decisive one. While similar activities around the Mediterranean had focused on identifying fossilized remains in abandoned areas, their work concerned traces of ancient occupation on land which was still being occupied and exploited. In England, the contrast was particularly strong due to the profound transformation of agricultural techniques in the late Middle Ages. The development of the wool trade led to an explosion of livestock rearing and the cultivation of fodder crops, so that vast swathes of arable land were put over to pasture from the 17th to 19th centuries. The previous open-field system gave way to enclosures, larger unified tracts of land separated by hedges. This was the pattern observed by aerial surveyors in the early 20th century. The absence of mechanical agricultural techniques contributed to the preservation of traces of medieval fields, particularly in cases where earthworks were present: ridge-and-furrow systems, lynchets, etc. The traces of old systems of land usage can sometimes be seen under modern fields, as shown in the striking photograph in Figure 2.4.

**Figure 2.4.** *Example of a medieval ridge-and-furrow system underneath the present enclosed landscape, near Didbrook, Gloucestershire, England (photograph by Philip Halling for The Geograph project collection, source: wikimedia.org/wikipedia/commons). For a color version of this figure, see www.iste.co.uk/robert/resilience.zip*

### 2.1.2. *The allure of a new vision*

In France, the work of English surveyors, published in the *Geographical Journal*, was disseminated by the geographer L. Aufrère in the *Annales de géographie* (Aufrère 1935). Aerial photography was also adopted in other scientific disciplines, with researchers drawing attention to the heuristic value of these documents compared to cartography. A summary of the contributions made by this technology in different domains was published in 1948 in *La découverte aérienne du monde*, edited by the sociologist Paul-Henri Chombart de Lauwe. In the book, the geographer Pierre Marthelot speaks of "new pages in the 'planetary archive'" (Chombart de Lauwe 1948, p. 98). Photographs were presented as a more objective medium than maps, since in the former case the representation is not generalized in any way. Aerial photography possessed an "authentic character", with a "synthetic value that no map will ever be able to match" (Chombart de Lauwe 1948, p. 281). For the ethnologist Marcel Griaule, the objective and concrete nature of photographs means that they are a more "direct" source than maps. A photograph will clearly show, for example, the system of relations connecting man, the plant kingdom and the animal kingdom. Griaule was particularly struck by the capacity of photographs to project past and present together in a single view:

> All of this bygone dynamic is thus projected into a single view, fixed in a single picture [...]. The image remains present to the eye, or at

least in the picture, offering a rich, direct and immediately interpretable representation of human implantation on the terrain[2]. (Griaule 1948, p. 182)

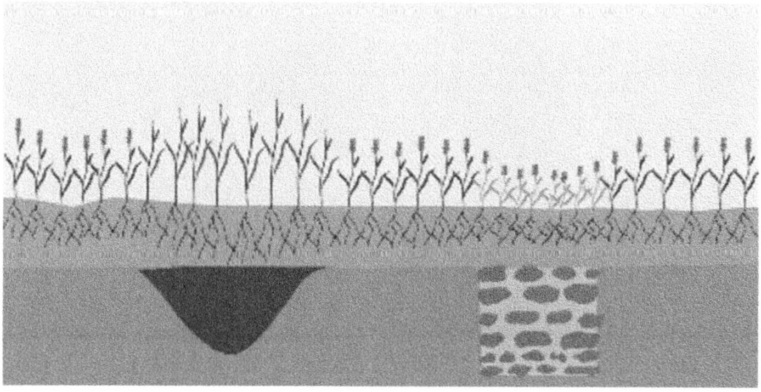

**Figure 2.5.** *Plants located directly above archeological remains receive more or fewer nutrients and grow differently to their neighbors, resulting in different colors and/or heights. Wheat planted over a ditch structure will stay green for longer, for example, or will ripen faster above a stone structure (Robert, 2019). For a color version of this figure, see www.iste.co.uk/robert/resilience.zip*

The state of the ground surface recorded in an aerial view is much richer than that found on a map, as it is not the result of a selection process. Objects are detected primarily through their spectral signature, based on their physical components, printed onto photo paper. Each object receives solar radiation and diffuses it according to its own characteristics and the wavelength of light. Information is given by a range of tones, varying as a function of the quality and nature of light received. Contrasts between different materials make it possible to distinguish the boundaries of different structures, visible at ground level. Variations, showing traces of former activity, are particularly evident on bare earth or in the form and color of vegetation; plants grow differently depending on the available quantity of organic materials, minerals, and the humidity present in the substrate. The presence of archeological remains results in local transformations in the soil content and in water retention (Chevallier 1971; Scollar *et al.* 1990) (Figure 2.5). Crawford (1929) established precise descriptions of physical conditions, which can betray the presence of archeological remains when seen from above. Throughout the 20th century, further contributions were made by aerial surveyors, who codified types in a way which resulted in a veritable "language of traces". Many handbooks and collections of aerial

2 *Toute cette dynamique du passé se projette donc en une seule vision, se fixe sur un seul cliché [...]. L'image demeure dans l'œil, ou celle du cliché, offre une représentation riche, directement parlante, immédiatement lisible des travaux humains d'accrochement au sol.*

photographs used in archeology devote a whole chapter to this subject (such as Scollar *et al.* 1990, pp. 33–77).

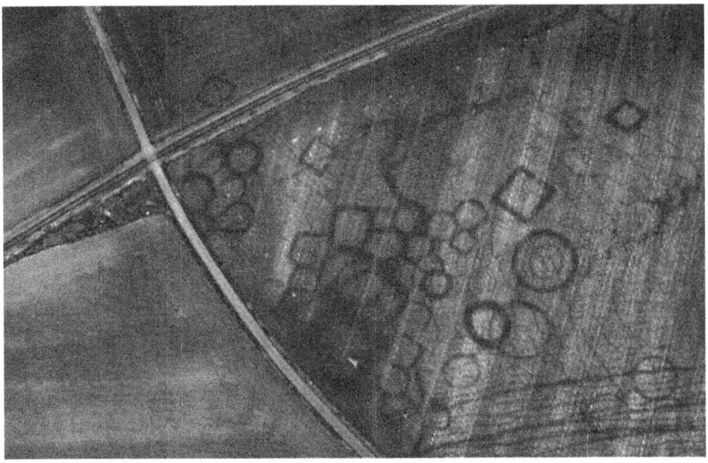

**Figure 2.6.** *Crop marks showing traces of funerary structures: in this case, cremation tombs in the Gallic necropolis at Grézac (Charente-Maritime, France). Photographed by the aerial surveyor Jacques Dassié (source: wikimedia.org/wikipedia/commons). For a color version of this figure, see www.iste.co.uk/robert/resilience.zip*

Aerial surveyors make a distinction between crop marks (phytographic indicators), which correspond to contrasts in the growth and color of plants (Figure 2.6), soil marks (pedographic indicators), consisting of modifications in the color or nature of soils[3], or shadow marks (sciographic indicators) where microreliefs are revealed using horizontal light sources at the ground level. This final technique is used in combination with other indicators, and is particularly suitable for low-altitude oblique aerial photography, as the tilt of the airplane can be adjusted so that the lighting highlights even the slightest variations in relief. Vertical aerial photographs are less suited to revealing marks of this type, but they are particularly valuable for identifying damp marks (hydrographic indicators) on bare soil during the winter. When the weather suddenly becomes more clement after a long rainy period, the surface of the soil dries, with more humid zones indicating the presence of previously occupied sites (Robert 2011, pp. 73–98).

As the detection of objects is largely determined by the material conditions in which a photograph is taken (humidity levels, vegetation type, etc.), different

3 For example, Roger Agache photographed plans of ancient villas, visible in the form of light-colored limestone fragments, turned up by plowing, on top of darker soil in the Somme region of France (Agache 1978).

temporalities collide. Traces generated by forms not seen above ground for centuries can be revealed by photography, as they continue to impact the present landscape by means of their effects on soil granularity, vegetation growth, etc. From a material perspective, these traces of the past thus belong to the present. Geographers, archeologists, or geologists may choose to note the presence of traces by means of classification: a categorization which is not automatically present in the photograph. From the 1940s on, interpreters increasingly tended to treat the forms of past constructions in a different way to present forms.

### 2.1.3. *Are we moving toward erasure of the past?*

Thus, in *La découverte aérienne du monde*, past and present traces are discussed in two distinct chapters: *La vision aérienne et les civilisations disparues* (Aerial Vision and Disappeared Civilizations: P.H. Chombart de Lauwe) and *Les civilisations vivantes. Les villes et leurs structures* (Living Civilizations. Towns and their Structures: M. Parent). In the former, monuments still visible on the surface are ignored in favor of traces of past drainage systems or field boundaries, presented as "long-abandoned systems" which can now be seen "beneath the present field arrangement" (Chombart de Lauwe 1948, pp. 249–250):

> Thus, the field of aerial vision extends to dead civilizations. The Earth's surface is not solely the canvas for the present constructions of living men. By examining it closely, we can detect the marks left by men in the earliest stages of our History, as if through tracing paper. By examining aerial documents, we become increasingly aware of successive layers left by successive civilizations, some of which depended on the others. The boundaries of ancient land divisions, corresponding to a different social system, are visible beneath more recent agricultural structures[4]. (Chombart de Lauwe 1948, p. 277)

According to P.H. Chombart de Lauwe, there are thus two possible ways of reading aerial photographs: one focused on the forms reflecting present society and the other focused on the traces of former systems, reflections of bygone societies

4 *Nous voyons donc s'étendre jusqu'aux civilisations mortes le champ de la vision aérienne. La surface de la terre n'est pas seulement le support des constructions des hommes d'aujourd'hui. En l'examinant attentivement nous voyons réapparaître, comme recouvertes par un papier transparent, les marques d'hommes des premiers âges de l'Histoire. Nous prenons mieux conscience, dans cet examen des documents aériens, des couches successives de civilisations qui se sont succédé et dont les unes sont tributaires des autres. Sous les structures agricoles récentes nous distinguons les limites d'anciens partages correspondant à un autre système social.*

that are no longer considered part of the present. Where morphological analysis of maps highlighted the continuity of forms from past to present, the approach used to interpret photographic materials in the social sciences in the period following the Second World War tended to see the past and the present as distinct and separate entities. Notions of substitution and erasure were particularly prevalent, leading to an understanding of landscape as a series of different layers. This is particularly evident in the enthusiasm of aerial surveyors for the palimpsest model from the 1950s on.

## 2.2. A stratified landscape

### 2.2.1. *Erasure in the palimpsest model*

At the turn of the 20th century, the palimpsest metaphor, applied to landscape, was seen as an accumulation of surviving elements, more or less concealed in the forms of the present (section 1.1.3). From the 1920s to 1930s on, historians and archeologists came to favor a different interpretation, closer to the original meaning of the word, where old text was erased in order to make way for the new. The development of aerial photography seems to have played a decisive role in this semantic development; the new technology was compared to chemical methods used in palaeography, in the way it enabled researchers to read the landscape-palimpsest.

The palimpsest metaphor was first applied to landscape by historians and geomorphologists (section 1.1.3); now, however, the innovation is generally attributed to the first aerial surveyors. The American geologists E.B. Guttmann-Bond and S.J. Kluiving, for example, consider that aerial photography accelerated understanding of palimpsest landscapes (Guttmann-Bond and Kluiving 2015, p. 13). For the French geographer Marie-Claire Robic, "the power of the unseen: the emergence of forms and the palimpsest" was one of the major discoveries in aerial photography[5]; the author emphasizes the idea that discoveries "emerge", a notion present in expressions such as "shown up", "came out" and "made out", used with respect to the detection of elements (Robic 2012, p. 137)[6].

The archeologist Gavin Lucas considers O.G.S. Crawford to have been the first to use the palimpsest concept systematically in the context of landscape (Lucas 2012,

5 See (Besse 2009, p. 96).

6 This idea of emergence may be compared to the way in which the first morphologists were able to detect forms on large-scale maps a few decades earlier (section 1.1.1).

p. 115). In a chapter entitled *Deciphering the Palimpsest: Roman Roads* in his book *Archaeology in the Field*, Crawford writes:

> The surface of England is a palimpsest, a document that has been written on and erased over and over again; and it is the business of the field archaeologist to decipher it. The features concerned are of course the roads and field boundaries, the woods, the farms and other habitations, and all the other products of human labor; these are the letters and words inscribed on the land. But it is not easy to read them because, whereas the vellum document was seldom wiped clean more than once or twice, the land has been subjected to continual change throughout the ages. (Crawford 1953, p. 51)

According to Crawford, the landscape features shown on maps at the time of writing generally dated from the late 18th and 19th centuries, when the medieval field system was "swept away" and replaced by enclosures (Crawford 1953, p. 51). Crawford takes a very similar approach to Maitland and Randall in his approach to landscape as writing; however, unlike his predecessors, he emphasized the idea of erasure and change above that of inertia (Figure 2.7).

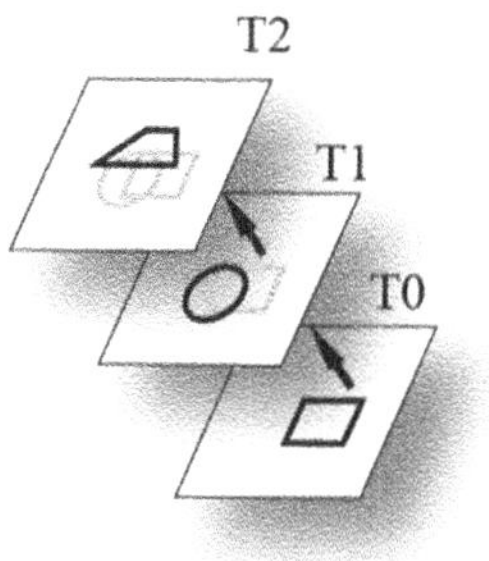


**Figure 2.7.** *Erasure in the palimpsest metaphor. Beginning in the second half of the 20th century, the notion of landscape as a palimpsest was used to emphasize ideas of erasure or transparency, rather than an accumulation of layers (Robert, 2019)*

In France, L. Aufrère, who had used the notion of palimpsest in geomorphology (section 1.1.3), chose to describe the agrarian systems discovered by aerial surveyors in the British Isles in the same way. In 1935, writing on the subject of the collections published by O.G.S. Crawford, he noted: "A whole palimpsest is coming to light, on which is written a history that seemed to have disappeared without a trace from the Earth's surface"[7] (Aufrère 1935, p. 390).

7 *C'est tout un palimpseste qui se révèle et où se trouve écrite toute une histoire qui semblait n'avoir laissé aucune trace sur la terre.*

In the 1950s–1970s, aerial photography was used in historical geography to look for deserted villages and for medieval land usage patterns abandoned to make way for enclosures (Beresford and Hurst 1971). William George Hoskins borrowed Crawford's palimpsest metaphor in his *The Making of the English Landscape*, contributing in no small way to the dissemination of the idea (Johnson 2008). The palimpsest is a key notion in the field of landscape archaeology, which emerged in the 1970s, combining different research traditions with the imperatives inherent in the emerging domain of preventive archeology. In a book considered to be one of the founding texts of landscape archeology, Michael Aston and Trevor Rowley highlighted the notions of erasure and burial, while also emphasizing the coexistence of objects from multiple temporalities within the present landscape:

> The landscape is a palimpsest on to which each generation inscribes its own impressions and removes some of the marks of earlier generations. Constructions of one age are often overlain, modified or erased by the work of another. The present patchwork nature of settlement and patterns of agriculture has evolved as a result of thousands of years of human endeavor, producing a landscape which possesses not only a beauty associated with long and slow development, but an inexhaustible store of information about many kinds of human activities in the past. (Aston and Rowley 1974, pp. 14–15)

The landscape-palimpsest idea was popularized in France by R. Chevallier (1929–2004) who, from the 1970s on, aimed to promote the English approach to landscape archeology in his own country. Chevallier is also considered to be the pioneer of archeological aerial photography in France. Like O.G.S. Crawford, his first experiences of aerial photography occurred in a military setting. Alongside his role as professor of Roman literature and civilization, Chevallier was also a colonel in the Air Reserve in the years following the Second World War; he was also president of the *Société française de photogrammétrie et photo-interprétation* from 1976 to 1980 (Poignault 2007). In 1953, he began a systematic examination of a complete aerial survey of Tunisia, carried out by the National Geographic Institute between 1947 and 1951 in concertation with the topography and photogrammetry service of the IGN[8]. According to G. Chouquer, IGN officials recruited Chevallier in the context of updates to the 1:50,000 map due to their own difficulties in distinguishing between contemporary landscape elements and traces of ancient land divisions, still clearly visible on the photographs (Chouquer 1990, p. 215). The topographic map was to feature only elements that still had a present function, while the reconstructed cadastral plans were published separately as the *Atlas des centuriations romaines de Tunisie* (Chevallier 1956). In an aerial photography handbook published in 1971,

---

8 *Institut national de l'information géographique et forestière*, the French national authority for geography, forest management and cartography.

R. Chevallier (1971) stated that "the Earth is a veritable palimpsest recording the history of humanity" (p. 168); the same image appeared in his manifesto for landscape archeology, *Le paysage palimpseste de l'histoire : pour une archéologie du paysage*. However, while the word "palimpsest" is present in the book's title, it was never precisely defined in the text. Instead, Chevallier highlighted the notion of a "fossilized" landscape, the intended object of the new landscape archeology:

> In this sense, landscape archeology aims to reconstruct elements which have disappeared, are fossilized, no longer have an evident function, but in some cases, may date from only yesterday. Our object here is to develop a way of reading this historical dimension from a landscape state, observed directly (on the ground or from an airplane) or known indirectly (from maps or written descriptions)[9]. (Chevallier 1976, p. 504)

### 2.2.2. *The fossil landscape*

The notion of "fossil" landscapes first emerged in the 1960s and, like that of the palimpsest, is closely linked to photographic interpretation. One of the first authors to use the term in France was the geographer Étienne Juillard. Writing in 1961, on the subject of English landscapes, he noted that the ancient landscape appeared as a veritable "fossil" (Juillard 1961, p. 993); Juillard's use of quotation marks indicates that the expression was not yet widely used in geographic contexts. The expression "centuriations, or fossilized Roman cadastral plans" appeared in the published papers of the first conference on aerial archeology, directed by R. Chevallier, and in Chevallier's own handbook of aerial photography (Chevallier 1964, 1971). However, Chevallier did not fully define the text until later (Chevallier 1976). Fossil landscapes were also a significant theme in the conference on landscape archeology, which he organized in 1977; they featured in the opening address given by the geographer Maurice Le Lannou, who referred to the fossil landscape as "a brand new notion", still rarely used by geographers (Le Lannou 1978, p. 3). Le Lannou also spoke of "relict" landscapes, a term borrowed from geomorphology, where it is used to describe terrain resulting from prior pedological developments, different from present conditions. English surveyors adopted the term "relict" in the 1970s to describe elements such as churches or wayside crosses "incorporated" into more recent village layouts (Aston and Rowley 1974, p. 123 and 140).

---

9 *En ce sens l'archéologie du paysage vise à reconstituer ce qui a disparu, est fossile, n'est plus consciemment fonctionnel, mais dans certains cas peut dater d'hier. Notre propos est de savoir comment lire dans un état de paysage observé directement (au sol ou d'avion) ou connu indirectement (par la cartographie, l'iconographie, les descriptions écrites) cette dimension historique.*

M. Le Lannou's text highlights the questions raised by different approaches to landscape in the second half of the 20th century. These approaches are not only tools for description, but also reflect societal choices and notions of social progress. In Le Lannou's view, a fossil or relict landscape is one containing features of no present utility, demonstrating "a transformation of a way of life or of an economy which has yet to be fully integrated in the geographical setting". The fossil landscape does not, therefore, correspond to contemporary lifestyles and indicates a "divorce, or better still a discordance, between the habitat and the economic choices – or imperatives – of the present" (Le Lannou 1978, pp. 3–4). The author cited the example of bocage areas in Brittany, which were then being cleared to facilitate land consolidation. In Le Lannou's speech, the fossil landscape appears as a form of resistance to the deep-reaching restructuring process operating in the countryside in the decades following the Second World War (land consolidation, mechanization of farming, hedge clearances, etc.). The author noted:

> [...] a growing opposition between man, the inhabitant, slow to envisage alterations to the traditional organization of the natural milieu, and man, the producer, subject to new imperatives, which often emerge rapidly and are seen as a law imposed from without[10]. (Le Lannou 1978, p. 5 and 7)

A feature can only be considered as a fossil with respect to an established norm. In the late 1970s, the type of agrarian landscape that constituted this norm was also a matter of debate. The geographer Pierre Georges, in *L'ère des techniques,* referred to inherited strata in a landscape as *décors d'autrefois*, the "scenery of the past", entrenched in "the resistance of small, outdated economies and aging populations, left behind by the train of modernization" (Georges 1974, p. 22). To this, M. Le Lannou responded:

> Dividing the landscape into dead and alive in this way betrays a reliance on an ethics of production, renouncing the eternal order of the fields which is at the heart of our European civilization of populations, farmers and artisans[11]. (Le Lannou 1978, p. 11)

10 *...une opposition croissante entre un homme-habitant lent à imaginer des retouches à son organisation traditionnelle du cadre naturel, et un homme-producteur qui introduit des impératifs inédits, le plus souvent abrupts et proprement subis comme une loi imposée de l'extérieur...*

11 *Pour couper ainsi le paysage entre la mort et la vie, il faut avoir choisi l'éthique de la production et répudié cet ordre éternel des champs qui fut au cœur de notre civilisation européenne d'habitants, de paysans et d'artisans.*

Thus, in the 1960s–1970s, the notion of the fossil landscape was inherently linked to social tensions over economic choices, notably in terms of agricultural production. The theme of landscape provided a theater for conflict between opposing ideologies. In this sense, the term "fossil" can be ambiguous: used in photographic interpretation to denote a trace or mark[12], it is also used in common parlance to suggest that something is old-fashioned or backward[13]. M. Le Lannou himself noted that if the expression "fossil landscape" was used indiscriminately, it could do more harm than good (Le Lannou 1978, p. 3).

While the notion of fossil or relict landscapes was a subject of debate in the 1970s, connected to broader discussions on the role of past forms in the present, by the 1990s it had passed into more general use, and in 1992 cultural landscapes featured in the UNESCO list of sites to protect. Fossil landscapes were specifically included as a sub-category of "organically evolved landscapes". This category includes "continuing landscapes which retain an active social role in contemporary society", and "relict (or fossil) landscapes [...] in which an evolutionary process came to an end at some time in the past, either abruptly or over a period". The UNESCO reference document goes on to specify that "significant distinguishing features are, however, still visible in material form" in these landscapes[14]; this definition is not dissimilar to those used by English surveyors and by M. Le Lannou 20 years earlier. However, for archeologists and historians, the term "fossil landscape" has come to refer exclusively to landscapes that have disappeared, buried beneath more recent elements.

### 2.2.3. *Buried landscapes*

In a handbook for historians published in 1995, A. Verhulst proposed a three-part classification of elements in a landscape:

– "functional elements", which retain their original function;

– "archaic elements", which have partially lost this original function;

– "fossil elements", which have lost all of their original function.

The second and third categories are conserved in the form of remains, features or traces, sometimes referred to as "relicts", and fall into the broader category of

12 The primary meaning of the word denotes remains or prints of plants or animals, buried in ancient sedimentary layers and generally preserved by petrification.

13 http://www.cnrtl.fr/definition/fossile, consulted 1st August 2018.

14 "Cultural Landscape". UNESCO World Heritage Center. http://whc.unesco.org/en/culturallandscape/, retrieved 7 June 2021.

monumental sources. Based on this categorization, Verhulst defined three main landscape types:

– The "current landscape", made up of elements of which the majority have a function in the context of rural life, and which include or are included in, overlap with or conceal older elements.

– The "former or archaic" landscape, which is "concealed beneath the many new marks which have been superimposed, so to speak, on it, for example by amalgamation". Thus, the current landscape only retains some elements of the former landscape, which have lost their function and are like "residual stains on the modern landscape".

– The "fossil or fossilized landscape", concealed or erased to an even greater extent, which "no longer bears any relation to the present landscape, under which it may only be preserved in a buried state, for example under a green blanket of fields, woodland or dunes…" (Verhulst 1995, pp. 10–11).

Like M. Le Lannou, who held that when older features in a landscape "cease to live, then they are no longer the object of geography, but rather of archeology" (Le Lannou 1978, p. 8), Verhulst indicates that fossil landscapes can only be detected using specific techniques, such as aerial photography. Each landscape type (current, archaic or fossil) has a corresponding set of "specific investigation, observation, interpretation and reconstruction techniques" (Verhulst 1995, p. 11). In practice, while aerial photographs appear to be a richer source than maps for studying the relations between, for example, natural, anthropic, past or present elements, the approach to these photographs differs according to the discipline in question. For instance, geographers will focus on present man-made forms, archeologists on past forms, geomorphologists on natural forms, etc.

Buried elements thus become the preserve of archeology[15], a domain which became increasingly specialized from the 1970s onwards with the development of rescue archeology. This consists of identifying and processing past remains prior to building or restructuring work. In Europe, rescue archeology has its origins in civil objections to the destruction of archeological remains in the course of post-war development (urbanization, material quarrying, land consolidation, rail and road infrastructures, etc.). In rural settings, the developments of rescue archeology and landscape archeology are closely linked; landscape archeology is a source of surveying techniques, notably aerial photography, map studies, toponymic and historical data analysis, studies of vegetation, collection of oral testimonies, etc. The 1970s saw an upsurge in public awareness of threats to the landscape, both as a historical source and as heritage. Aston and Rowley considered the landscape as an

15 On the subject of archeology as a "science of the buried", see (Boissinot 2011) (in French).

archive in its own right, and one which should not be left to decay on the grounds that it cannot be housed in a library or archive. Their research, particularly in conjunction with the development of the M40 motorway north-west of London, launched a new wave of interest in archeological sites as elements in a broader environment. Their approach aims to take account of landscape elements, networks of different sites, land organization systems, ancient paths and road networks, etc. (Aston and Rowley 1974, p. 14 and 173).

French researchers were slower to adopt low-altitude oblique aerial surveying techniques than their British counterparts, but by the 1960s–1970s, this approach played an important role in heritage awareness. Aerial surveyors began to discover more and more archeological sites concealed beneath modern field systems, while also noting the extent of destruction brought about by agricultural mechanization (deep ploughing, leveling, hedge clearance, etc.) and infrastructure development. In 1973, a special edition of *Archaeologia* magazine included remarkable aerial photographs from all over France, alongside articles intended to raise awareness of the "unacceptable" rate at which archaeological sites were being destroyed (Archeologia 1973)[16]. As in Britain, aerial photography provided a significant contribution to the development of heritage registers. Low-altitude oblique aerial surveying was generally used to the detriment of vertical photography techniques. While the latter are valuable for the detection and interpretation of former land division systems and road networks, they are less effective for close inspection of archeological sites. Using oblique techniques, the moment and angle at which a photograph is taken can be adapted to limit shadow effects, which hinder photogrammetric analysis. The visual cues used to detect the presence of archeological sites are harder to see in vertical aerial photographs (section 2.1.2). While high-altitude techniques tend to show persisting elements within contemporary landscape forms, low-altitude aerial photography, generally used to detect abandoned sites, focuses on the detection of traces that have been "erased". The systematic replacement of vertical aerial photography by oblique low-altitude techniques reinforces the idea that the ancient landscape is preserved under, and not within, the modern landscape. Along with the notions of palimpsest and fossil landscapes, this development contributed to the emergence of a strata-based understanding of landscape (Chouquer 2000, p. 26).

### 2.2.4. *The stratified landscape*

The notion of strata is used in geology to define deposits, originally sedimentary rock found on the sea floor or lake beds, which form coherent layers and remain in

16 See notably Louis Faton: "*Nous croyons au ciel*" p. 8-11, and "*La destruction de notre patrimoine archéologique n'est pas fatale*" by the architect Paul Dufournet, pp. 113–122.

the order in which they were deposited, except in cases of subsidence or folding (Derruau 1996, p. 34). The Danish scientist Niels Steensen – anglicized as Nicolas Steno (1638–1686) – established what is now known as the superposition principle, whereby layers should be horizontal, and the most recent deposits should be found at the top. Stratigraphy has been used in archeology since the 19th century. The order of layers means that they can be dated with respect to one another, and the presence of datable elements within one stratum means that the dates of other strata can be determined in a relative manner (Desachy 2008, pp. 24–42). In France, the idea of using the stratigraphic method in the context of landscape seems to have taken root in the years following the First World War, at around the same time as the introduction of aerial surveying in Europe. One of the first instances of relative dating of elements in a landscape is found in an article by L. Aufrère, published in 1935, concerning the work of English aerial surveyors: "A path may be dated in relation to a field or a bank. If one of these elements provides a chronological indication, then the whole agricultural landscape can be dated" (Aufrère 1935, p. 391).

The application of relative chronology to landscape – only visible in two dimensions on aerial photographs, in contrast with the three-dimensional structure of an archeological trench – was systematized by R. Chevallier. His 1971 handbook included a chapter on the temporal dimension of aerial photography, which he considered as a means of obtaining a synthetic vision, giving an overall impression of a whole time period, and moving beyond what H. Bergson called "the infirmity of our perception, condemned to view the film of a lifetime one image at a time". In Chevallier's view, photography was like "a stationary image of time" which could be read as a "chronological cross-section" (Chevallier 1971, p. 104 and 106). During the first French conference on aerial archeology, Chevallier called for the creation of relative chronologies based on reciprocal relationships between two or more landscape elements, "transformed into absolute chronologies by reference to historical or archeological data…" (Chevallier 1964, p. 15).

Chevallier held that landscapes contained the same types of superposition relationships as those used in geology, implying chronological successions. In *Le paysage palimpseste de l'histoire*, he stated that, in a landscape, time can be reduced to a two-dimensional space. Relative chronologies can then be used to establish logical temporal relationships between key elements:

> Drawing on analytical methods used in geology, we have shown that, in the best cases, it is possible, by analyzing the relations between these three elements [roads, plot layouts and habitat], to establish *relative chronologies* of the various states of a single landscape, including up to 9 or 10 strata. These relative chronologies may be converted into *absolute chronologies* by the insertion of dated

> elements. This is one of the keys to photographic interpretation[17]. (Chevallier 1976, p. 508, italics original)

In his handbook on photographic interpretation, Chevallier explains how the creation of a chain of causality allows us to work back from objects to their causes, based on the idea that if one phenomenon causes another phenomenon, then the former necessarily occurred first:

> The criterion of anteriority is intimately linked to that of causality, as causes occur before, or at the same time as, their consequences. Taking two phenomena which are connected by a certain function, the earlier of the two may be considered to have caused the other, and thus, logically, the morphological relations between structures must correspond to relations between the phenomena resulting in these forms. In the case of superposed forms of similar morphology, we must consider the possibility of there being an earlier cause; for example, a given parcel system or habitat distribution may result simultaneously from the same geological structure. In this way, objects are linked back to their first causes, known or otherwise[18]. (Chevallier 1971, p. 107)

An arrangement of houses in concentric circles, for example, may be explained by the existence of an ancient amphitheater on the site, the existence of which is deducted from the arrangement of the houses. The relations at work here are ones of simple causality: one cause corresponds to one, or a series of clearly identifiable effects; thus, predictions can even be made concerning future developments (Chevallier 1971, p. 109). Following Chevallier, several researchers attempted to establish a logical framework for the interpretation of landscape forms. Bernard Liger suggested that the stratigraphic nature of an observed landscape should be verified by checking "essential axioms"; for example, "if a network conceals another network, even partially, then the

17 *Nous inspirant des méthodes d'analyse des géologues nous avons montré que, dans les cas les plus favorables, il est possible, en analysant les rapports de ces trois éléments [voies, parcellaires et habitat], d'établir des chronologies relatives des divers états d'un même paysage, comportant jusqu'à 9 ou 10 strates. L'insertion d'éléments datés dans ces chronologies relatives peut permettre de les transformer en chronologies absolues. C'est là un des nœuds de la photo-interprétation.*

18 *Le critère d'antériorité est étroitement lié à celui de causalité, car leurs causes sont antérieures à leurs conséquences ou simultanées. De deux phénomènes liés par une certaine fonction, celui qui est antérieur peut passer pour la cause de l'autre et l'on supposera que les rapports morphologiques entre les structures doivent correspondre à des rapports entre les phénomènes provoquant ces formes ; si donc l'on a affaire à des formes superposées et de morphologie analogue, on postulera une cause antérieure, par exemple telle allure de parcellaire et tel groupement de l'habitat peuvent être déterminés simultanément par telle structure géologique. L'on remontera ainsi des objets à leurs causes, connues ou non.*

latter must predate the former" (Liger 1972, pp. 9–10). This analytical approach obeys a Cartesian logic, according to which a phenomenon may be understood by breaking it down into smaller units. From this perspective, a landscape is considered to be a sum of simple objects, an accumulation of entities that may be taken in isolation and interpreted separately in order to recompose groups at a higher level. In the early 1990s, G. Chouquer suggested that relative chronologies might be established "by breaking down the structure to the greatest possible extent" (Chouquer 1990, volume I, p. 293). Once again, we see that landscape may be read as a language. Max Guy and Michel Passelac described the approach as follows:

> Images are read using a very simple formalism of the grammatical type: characters, such as 'parcel boundaries' or 'constructions' are linked to create elementary forms, just as letters may be assembled into words. The relations between these forms may then be interpreted as words which combine into phrases[19]. (Guy and Passelac 1991, p. 117)

Using a conventional approach to photographic interpretation (applied, in this case, to the new remote sensing techniques developed in the 1980s), Guy and Passelac expressed the idea that any boundary which "takes account" of the existence of another boundary must post-date that boundary. Boundaries with few or no subordinate limits, which appear to be "laid onto the landscape", are considered to be "recent objects, later in date than most of the elements in the landscape". Subordinate lines are interpreted as recent secondary divisions; an initial structure, subjected to successive land divisions (e.g. for inheritance purposes) or land consolidations as it aged, may be reconstructed by removing these secondary lines (Guy and Passelac 1991, p. 117 and 120) (Figure 2.8). This idea is connected to historical research aimed at reconstructing original open field systems. In the 1960s, historians and geographers believed that they had found a simple way of identifying temporal dimensions in a landscape, using the notion of divisions or *gewann*. Modern or contemporary documents indicated the boundaries of original divisions of open field systems, progressively divided into smaller units as the land was shared along multiple heirs (Verhulst 1995, p. 24). Using the *Rückschreibung* method (literally "writing backwards"), based on the same principles as regressive analysis, German historians and geographers believed that the original units could be reconstructed, working back from the cadastral plans of the 19th century and through parcel ownership documents from the 18th, 17th and even 16th centuries. Morphologists were thus able to establish a relative chronology of open field boundaries and strip farming divisions (Figure 2.8).

---

19 *...la lecture de l'image fait appel à un formalisme très simple de type grammatical : des caractères, par exemple "les limites de parcelles" ou "des constructions", seront reliés pour constituer des formes élémentaires, comme les lettres sont assemblées en mots. Les relations de ces formes seront interprétées ensuite comme les mots forment des phrases.*

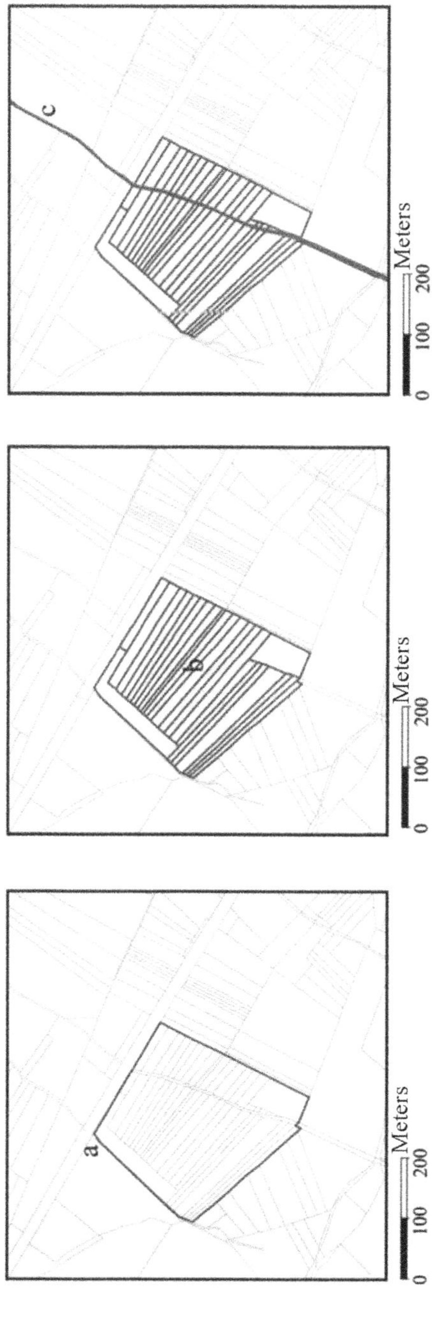


**Figure 2.8.** *Example of stratigraphic interpretation, applied to landscape forms. Parcel layout (b) is considered to post-date the initial open field (a). Road (c), which passes through both a and b without adapting to the forms of these layouts, is considered to be a later addition: a < b < c (Robert 2019). For a color version of this figure, see www.iste.co.uk/robert/resilience.zip*

## 2.3. The synchronic view

The stratigraphic approach to landscape interpretation implies a different view of time to those used previously. Where the diachronic temporality involves a continuous flux, mixing multiple times, the synchronic view involved in stratigraphy is one of cross-sections based on observing a state at a given time t.

### 2.3.1. *Synchronizing forms and society*

F. de Saussure developed the notion of synchrony in 1916, in parallel, and in contrast, to that of diachrony. He considered language as an organized whole: an "organism", making up a "system of signs expressing ideas", with its "own order" (de Saussure 1986, C. III, § 3). Comparing language to a game of chess, he noted that certain characteristics (such as the material from which the pieces are made) can be changed without altering the game, whereas if others (such as the number of pieces) are changed, the game is transformed into something else. Saussure thus made a distinction between internal elements, with the capacity to change the system, and external elements which are not themselves part of the system. In his view, history belonged to this latter category, as the evolution of words cannot explain present states of a language (de Saussure 1986, C. V). He proposed that, within each science, a distinction should be made between an axis of simultaneity, which "concerns relations between things that coexist, relations from which the passage of time is entirely excluded" (synchrony), and an axis of succession, along which "one may consider only one thing at a time", but where "we find all the things situated along the first axis, together with the changes they undergo" (diachrony) (de Saussure 1986, C. III, § 1). Synchrony thus cuts across time, allowing us to describe "states" of a language seen in isolation, while diachrony describes the phenomena by which a language passes from one state to another. Fernand Braudel (1958) later used the metaphor of a "descent down the slope of time" (*descente selon la pente du temps)* (p. 739).

In de Saussure's view, elements of disparate origins hold together at a time t as they belong to the same value system: "A language is a system of pure values, determined by nothing else apart from the temporary state of its constituent elements" (de Saussure 1986, C. III, § 1). In its general sense, the value describes the measurable character of an object as a thing which may be exchanged or desired; it naturally varies between groups, individuals and historical and geographic contexts. Value is intrinsically linked to the notion of utility, in concepts such as "exchange value" or "useful value", which correspond to "the quality of a good or a product in terms of its ability to respond to a given social need[20]" (Rey and

20 ...*qualité qu'a une marchandise ou un produit de satisfaire un besoin social donné.*

Rey-Debove 1992, p. 2061)[21]. A system at a time t thus consists of an assembly of elements within a set of shared values. For example, the historian Bernard Lepetit explains the association of multiple temporalities within the urban landscape by the fact that contemporary societies assign meaning to inherited forms. Towns and cities are thus made up of disparate fragments that have evolved at different rates and in different ways. Certain forms are still used, because of a series of partial adjustments to ensure that the building remains compatible with the activities it houses. For Lepetit, if adjustment ceases to be possible, the urban object will be transformed into a trace: "a past left behind by the present"[22]. The dividing line between past and present is thus established by contemporary social practices (Lepetit 1995, p. 292). The maintenance of different temporalities in a landscape thus involves a social aspect, in which inherited forms may be reinterpreted or assigned new meaning.

In the decades following the Second World War, discussion concerning the presence of past forms in the present was influenced by the economic and social transformations, which were then at work in French society. Legacy elements in landscapes were perceived as limiting factors in the articulation of a new, emerging society with a material environment which, it was feared, did not correspond. Pre-existing parcel layouts, in particular, were considered as a hindrance, and land consolidation was seen as a necessary step in any urban or rural modernization program. In texts published after 1945, these existing land divisions were considered to be a "chronic disease", limiting the evolution and modernization of towns. Architectural journals and brochures published by the Ministry of Reconstruction and Urban Development were careful to highlight the absurdity of reconstructing new habitats on top of "outdated traces" (*traces périmées*). The "unhindered disposition of land" (*libre disposition des sols*), through land consolidation and via the destruction of substandard habitations, seemed to offer a solution for planned developments of urban centers, in the "interest of the community" (*l'intérêt de la collectivité)* (Atelier d'architecture et d'urbanisme Treuttel Garcias Treuttel 1996, pp. 31–33).

The territorial development policies adopted in the period following the Second World War were heavily influenced by the progressist current (Choay 1965), which had its origins in the 19th century but only truly took form with the publication of the Athens Charter in 1942. This charter promotes functional urban development focused on meeting fundamental human needs. The focus on function means that this approach to urbanism implies two fundamental divisions: between nature and culture, and between past and present.

---

21 See also http://www.cnrtl.fr/definition/valeur.

22 *...un passé dont le présent s'est détaché.*

### 2.3.2. *The negation of nature and the division of past from present*

According to the progressist view, urban development must show man's dominance over nature. Thus, for Le Corbusier (2013): "A city! It is the grip of man on nature. It is a human action directed against nature, a human organism both for protection and for work [...]" (p. XXI). This "grip" was to be expressed in man's capacity to project geometric forms, bearing relation to the natural site, onto the land. This dissociation was considered to be a prerequisite for human liberty, as nature was seen as a constraint on creation: "When man is free, his tendency is towards pure geometry [...]. The things he makes for himself are a creation which contrasts all the more with his natural surroundings because its aim is closer to his mind, and further away and more detached from his body" (Le Corbusier 2013, p. 22). Alluding to the location of Paris with respect to a ford over the Seine River, the architect lamented the fact that "the pack-donkey's way is responsible for the plan of every continental city, including Paris, unfortunately" (Le Corbusier 2013, p. 6). While the non-adaptation of a town or city plan to the local topography might be seen as a matter of human liberty, it also marks a clear rupture with regard to its past trajectory and history.

For Le Corbusier and the architects of the CIAM (*Congrès internationaux d'architecture moderne*), the advent of the industrial era was an essential watershed; they noted an increasing mismatch between the form of the city and the evolution of technology, which they believed to be driven by its own, relentless force (Le Corbusier 2013, p. 39). The dynamics are found in society, which has its own history. Urban forms should thus be characterized by a series of breaks, marking these readjustments. Ildefons Cerdà, one of the first urban planners, proposing a new form for the city of Barcelona in 1859, believed that he could "restart" history, notably by eliminating the historic city center. In his view, inhabitants would reject the center in any case: "he who has known happiness will never desire a return to barbary"[23] (Cerdà 2018 [1859], p. 26). In a similar vein, Le Corbusier, in the "Voisin" plans of 1922-25, recommended that the historic districts of the Right Bank in Paris should be replaced completely (Le Corbusier 2013, pp. 278–288); later, in the Athens Charter, he rejected the notion of the reversibility of forms and the culturalist approach to urban planning:

> The practice of using styles of the past on aesthetic pretexts for new structures erected in historic areas has harmful consequences. Neither the continuation of such practices nor the introduction of such initiatives will be tolerated in any form. Such methods are contrary to the great lesson of history. Never has a return to the past been recorded, never has man retraced his own steps. (Le Corbusier 1973, § 70)

23 *...celui qui a connu le bonheur ne voudra jamais retourner à la barbarie.*

During the post-war period, society underwent major technical and economic transformations, just as it did in the late 19th century. For urbanists of the progressist school, the relation between form and function was one of strict reciprocity, based on synchronizing forms and society and using a system of values that held the past at arm's length. If the form, as "container", threatened to hinder the progression of the contents, then cutting off past from present became a major societal imperative.

Thus, in the decades following the Second World War, approaches to explaining the present focused less on the genesis of processes inherited from the past, and more on economic and social relationships, in a state of constant renewal. The organization of a space was seen as a result of social practices that may be modeled using rules, rules which the "new geography" aimed to highlight from the 1960s on (Haggett 1965). In terms of levels of observation, form was no longer considered to be sufficient as a means of expressing the organization of a space. The associated techniques of map analysis, aerial observation and morphological analysis were no longer seen as suitable tools for synthesis, as they emphasized the visible elements of a landscape – *le visible du paysage* – to the detriment of the intangible elements at work in a space: decision processes, the dissemination of information, etc. (Dauphiné 1998, p. 57).

### 2.3.3. *The break between past and present in methodology*

If successive societies adjust their spatial forms to the present, then the former landscape will be erased, relegated to the domain of traces that can only be detected using specialized techniques. Methodological choices made in the context of rescue archeology reinforced this idea of archeology as a means of interpreting buried remains. While a wide variety of archeological survey techniques were used in the 1970s–1980s (field walking, aerial surveys, test pits, documentary analysis, etc.), systematic trench-cutting, using mechanical diggers, came to dominate the field by the mid-1990s. In this way, samples are taken from an area by removing the topsoil, in which any archeological remains are likely to have been destroyed by plowing. However, the removal of the ploughing layer in rural settings, or of modern and contemporary fillers in urban environments, means that the dynamics of traces up to and including the present day are ignored. Archeology is thus limited to the study of objects that are considered to have no function in present society, which only play a part in a contemporary system of values insofar as they are considered to constitute a significant aspect of our heritage. The majority of objects that make up a landscape – such as unmarked pathways – are thus left aside (Robert 2021).

The morphological analysis techniques used by historians and archeologists were also affected by the notion of a temporal break; this change can be seen in the debate surrounding the use of "overly recent" maps. For example, following the publication of their *Atlas des centuriations romaines de Tunisie,* R. Chevallier and A. Caillemer were forced to justify the fact that their photographic interpretation diagrams were overlaid onto maps featuring contemporary landscape features, such as railroads (Caillemer and Chevallier 1958). In 1983, the archeologist Alain Ferdière recommended:

> ... evidently, the elimination, first and foremost, of the most recent modifications to the plot layout (regroupings etc.), along with other very recent structures (roads, dwellings): otherwise, there is a risk of error, all too common in investigations of ancient cadastral plans, for example, where railroad lines or the boundaries of regrouped plots are used in the reasoning[24]. (Ferdière 1983, p. 162)

Regressive analysis involves a sorting process, beginning with the extraction of the most recent elements. In terms of transport networks, axes constructed from the 19th century onward, such as railroads, must be eliminating. For plot layouts, the reconfigurations of the post-war decades still constitute an important psychological break point. In the 1990s, plot layouts were analyzed based on documents from before the mechanization of agriculture; the resulting regroupings and amalgamations were seen as a watershed after which no information on earlier states may be found (Robert 2003b).

### 2.3.4. *The opposition of nature and culture*

It is our belief that the break imposed between past and present is inextricably linked to that between nature and culture: a strict separation between the things of nature (soil, vegetation, water) and those of culture (functional social objects, such as roads and buildings). This is evident in photographic interpretation practices, where archeological remains are catalogued on specialist maps, despite the continued presence of these traces in the present; the continued existence of the traces is the main thing that enables their detection by cameras or satellite captors. While past and present, societal and natural forms are intimately connected within Earth's complex landscapes, the categories used to read and interpret traces have a tendency to minimize, if not ignore, these relationships.

---

24 *[d'] éliminer en premier lieu, bien sûr, les modifications les plus modernes du parcellaire (remembré...) comme les autres structures (voies, habitats) très récentes : on s'expose sinon aux errements encore trop courants des recherches des cadastres antiques par exemple, qui utilisent dans leur raisonnement voies de chemin de fer ou limite de remembrement.*

First and foremost, the trace, as a "physical mark or material left by someone or something"[25] (Rey and Rey-Debove 1992, p. 1993)[26] has a material existence independently of any function, use or value. The role of photo-interpreters is to identify which of the many marks seen on an aerial image possess a function, i.e. a connection to a "given activity assigned to an element in a group, or to the group itself", notably in "specific activities associated with man"[27] (Rey and Rey-Debove 1992, p. 801)[28]. Elements in a landscape that possess a function generally also possess a status, assigned by a legal decision, a momentary compromise between the multiple values laid onto landscape elements over time. This is the reason why the place assigned to past forms on topographic maps has always been a matter of debate. The recruitment of historians (such as O.G.S. Crawford and R. Chevallier, sections 2.1.1 and 2.2.1) by mapping organizations, in a context of photographic interpretation, is discussed in this respect. Landscape elements that continue to have a function – such as churches – can be easily integrated into maps, where they are represented in the same way as present forms, using solid lines and the same font. Ruins and remains are more systematically included in cases where they possess a legal status or attract tourists, essentially taking on a new function. The representation of other traces of the past, however, is negotiated on an almost case-by-case basis using specific codes of indication. For example, the remains of a former vehicular carriageway, shown by a dashed line on 1922-type French ordnance maps, are not included on the 1993-type 1:25,000 map, where classification is based on viability (Bèzes *et al.* 2014). Discussions concerning the representation of ancient centuriation marks on updated maps of Tunisia were only resolved with the help of a historian (section 2.2.1), another illustration of the fact that these debates are more than just a matter of cartographic techniques.

A vast number of buried traces are not indicated on topographical maps, for lack of a function, status, meaning or exchange value with the capacity to "respond to a given social need"[29]. Traces of this kind are only found on specialized maps created by historians and archeologists, in the case of past remains, or by geologists and geomorphologists, in the case of distinctly environmental elements. If the physical dimension of forms (nature) is reintroduced, then the categories of "fossil" or "relict" landscape (section 2.2.2) become problematic: while the societies which created certain landscape forms are long gone, the forms continue to have a role in nature, whatever the value or status assigned to them by present societies (see the notion of ecological land-use legacies, presented in section 6.3).

25 ...*marque physique, matérielle laissée par quelqu'un ou quelque chose.*

26 See also http://www.cnrtl.fr/lexicographie/trace.

27 ...*activité déterminée dévolue à un élément d'un ensemble ou à l'ensemble lui-même... [...] activités spécifiques attachées à l'être humain.*

28 See also http://www.cnrtl.fr/definition/fonction.

29 http://www.cnrtl.fr/definition/valeur.

## 2.4. Conclusion

In the years which followed the Second World War, attention shifted from the persistence of landscape to its transformation. Using aerial imaging, a new technique which was quickly adopted in both geography and archeology, the palimpsest metaphor once again proved its relevance; however, where late 19th century researchers used the notion in the sense of accumulation, the new generation of specialists saw the palimpsest in terms of erasure. Aerial archeological surveying, carried out by eye and at low altitudes, focused on the identification of abandoned sites; this reinforced the notion that the landscapes of the past are buried beneath those of the present. Thus, they can only be discovered using specific techniques, in which the landscape is interpreted as a series of geological or archaeological strata. Taking a synchronous approach to time, landscape elements forming a system at a given time t were considered to form part of a single social value system in operation at that moment (Figure 2.9). The continued existence of ancient plot limits and paths was seen as a hindrance to future development, and these elements were not to be included in future plans. The progressist school of urban development promoted a clear break between present and past forms, which they did not consider to be compatible with post-war society. Architects and planners called for a "great leap forward", moving toward more "suitable" organizations of space; much emphasis was placed on planning approaches that favored new forms, cut off from the historic and geographic determinism of their location.

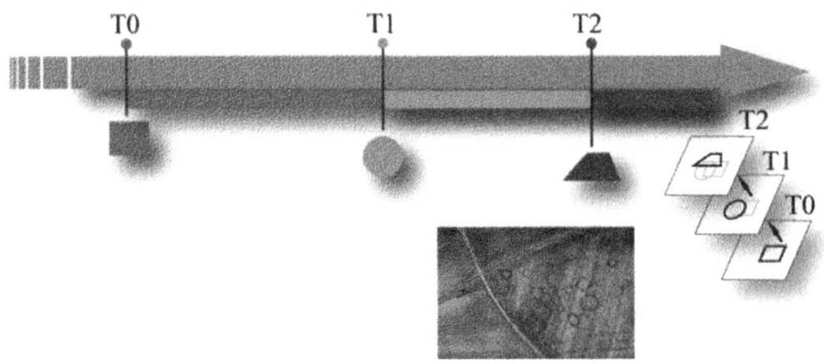


**Figure 2.9.** *Synchrony in landscape forms. Landscape is reorganized in a manner synchronous to societal transformations. Past forms can only be accessed using specific techniques (aerial imaging, removal of topsoil, etc.). Past landscapes are buried beneath present landscapes, with a point of reference in a constantly readjusted present (Robert, 2019, photograph by Jacques Dassié, source: wikimedia.org/wikipedia/commons). For a color version of this figure, see www.iste.co.uk/robert/resilience.zip*

In the course of the 1990s, the notion of the palimpsest, in the erasure sense, fell from grace as a means of describing the relations between past and present in a landscape. Geographers called for a re-inclusion of past traces in attempts to understand long-term spatial trajectories[30]. For Roger Brunet, landscape is more than a palimpsest, "insofar as these archeological traces are, in their turn, agents of new systems (if only as obstacles)"[31] (Brunet 1974, p. 123). The past forms part of the trajectory of systems and intrudes on the present, whether we like it or not. As the geographer Jean-Luc Piveteau put it, "landscape is not only a 'palimpsest', in the sense of juxtaposed or overlapping 'deposits' of different ages, but also a 'protrusion' of multiple, one might say truncated, times"[32] (Piveteau 1990).

The palimpsest metaphor also attracted criticism in the field of archeology. The English archeologist Geoff Bailey defined five different types of palimpsest, highlighting the complexity of temporal relations between different layers of a single stratigraphy (Bailey 2007, p. 203)[33]. In archeogeography, too, G. Chouquer noted that the notion was increasingly redundant: "the ancient spatial system is not, therefore, just a palimpsest, but also a form which determines – directs, marks out and draws, in part, the later landscape"[34] (Chouquer 1997a, p. 15). In this period, from the late 1990s into the early years of the new millennium, there was a shift in focus onto dynamic organizations, articulating past and present forms in complex trajectories, rather than synchronous sets of elements in a landscape associated with, and confined to, given societies. Among landscape researchers, too, the notion of the palimpsest was reassessed, with the aim of taking account of the complex dimension of heritage in a landscape[35].

---

30 In this book, the expression "long term" is used in the historical and archeological sense, indicating a duration of several centuries.

31 *...dans la mesure où ces traces archéologiques sont à leur tour des agents de nouveaux systèmes (même à titre d'obstacles).*

32 *...le paysage n'est pas seulement "palimpseste", c'est-à-dire "dépôts" juxtaposés ou chevauchants de différents âges, mais il est également "affleurement" de multiples durées, comme tronquées.*

33 See also Olivier (2009) and Lucas (2012).

34 *...le système spatial ancien n'est pas ainsi seulement un palimpseste, mais aussi une forme qui détermine, c'est-à-dire oriente, rythme et dessine en partie le paysage postérieur.*

35 See Corboz (1983).

# 3

# Landscape: The Articulation of Past, Present and Future

The 1990s saw a deep-reaching transformation in the perception of the dynamics at work in the landscape. The increasingly systematic pairing of landscape form analysis with data collected within the framework of rescue archeology highlighted the limitations of the regressive approach and of semiological landscape analysis. The meeting of the two disciplines resulted in a new vision of temporalities in landscape, as developments in cartography or aerial photography had contributed to a renewal of approaches in the late 19th and early 20th centuries.

## 3.1. The 1990s: a period of revival

### 3.1.1. *Learning from the ground*

In the 1990s–2000s, rescue archeology gained an institutional framework, notably in the wake of the European Convention on the Protection of the Archaeological Heritage (1992). In France, the law of January 17, 2001 established a system of charges intended to finance archeological work, and an administrative service, the Inrap (*Institut national de recherches archéologiques préventives*) was established to manage both archeological surveys and digs. The systematic test trench system, using mechanical diggers, enabled archeologists to analyze hundreds of hectares of land, giving access to information that may be precisely dated. Alternative, non-destructive techniques, such as aerial surveying and planimetric document analysis, were also used. In cases of large-scale developments or transects covering multiple kilometers, the results of morphological analysis were compared with archeological data. The three-volume *Les formes du paysage*, published between 1996 and 1997, provides an early summary of many of these developments (Chouquer (ed.) 1996a, 1996b, 1997b).

In the field of morphological analysis, the early 2000s saw a greater convergence between archeology and the Earth science disciplines (Van der Leeuw and Thiébault 2000), giving rise to a new discipline of geoarcheology, which takes account of the role played by human societies in land transformation processes. In this context, greater efforts were made to situate past plot layouts and road and path networks within their natural environment.

### 3.1.2. *Re-evaluating the nature/culture dichotomy*

Increasingly close collaborations between archeologists and environmental scientists (geomorphologists, climatologists, palynologists, etc.) emerged during the 1990s, notably in the context of studies focused on specific regions; these formed a testing-ground for multidisciplinary approaches[1]. The international, multidisciplinary *Archaeomedes* program, financed by the European Union from 1992 to 1998, played a particularly heuristic role in terms of discussions concerning the analysis of long-term interactions between societies and their environments. The aim of the program was to identify and highlight the anthropic and natural dynamics of soil degradation, notably desertification in the Mediterranean basin (Van der Leeuw 1998)[2]. In the French part of the program, geographers and archeologists worked together to analyze the mobility of rural habitat from protohistory to the present day in eight microregions of the Rhône Valley and the eastern Languedoc (*Archaeomedes* 1998). One result of this program was the adoption of the self-organization theory in archeology, which made a significant contribution to a resurgence in morphological landscape analysis from the 2000s (section 3.4).

The systemic approach played a key role in interdisciplinary connections, promoting an integrated approach to the relationships between societies and their environments. In 1978, the geographer Georges Bertrand put forward the concept of a "geosystem" (*géosystème*) as a tool for landscape analysis[3], designed to encompass

1 *Programme interdisciplinaire de recherches sur l'environnement* (PIREN) on major waterways, *Groupements d'intérêt public* (GIP Hydrosystèmes), the *Programme environnement, vie et sociétés* (PEVS), etc. See, for example Bravard and Magny (2002), Muxart *et al.* (2003), and Burnouf *et al.* (2009).

2 *Archaeomedes* I (1992–1994): *Understanding the Natural and Anthropogenic causes of soil degradation in the Mediterranean Basin* (Environment Program of the European Commission DGXII); *Archaeomedes* II (1996–1999): *Policy-relevant models of the natural and anthropogenic dynamics of degradation and desertification and their spatio-temporal manifestations,* Environment Program of the European Commission DGXII, coordinated by S. Van der Leeuw, Cambridge University.

3 Later developed into the GTP (*géosystème, territoire, paysage* – geosystem, territory, landscape) approach.

sets of relationships linking both social and ecological aspects (Bertrand 1978). In 1991, Bertrand once again called for an abandonment of the period-based approach to archeology and the adoption of a broader agrarian archeology, which would potentially form an "'interface' discipline between the natural and social sciences"[4] (Bertrand 1991, p. 17). From the 1990s on, the systemic approach led to an increase in multidisciplinary activity, reintroducing human and historical aspects into studies which were, initially, purely physical or ecological. The crossover between paleo-environmental and archeological studies often centers on the concept of the "anthroposystem", defined as "a system of interaction between two groups made up of one or more sociosystems and one or more natural and/or artificialized ecosystems, forming part of a given geographic space and evolving over time". This notion was intended for use in "studying the interactions between societies and natural environments by means of a long-term, systemic, integrated approach, with a view to sustainable development"[5] (Lévêque *et al.* 2003, p. 110). It was within this context that morphological analysis, applied to historic landscapes, underwent a renewal.

Discussions between archeologists, historians, geographers and environmental scientists resulted in the emergence of a specifically French school of morphology. Initially referred to as *archéomorphologie* or *morphologie dynamique*, the discipline finally came to be known as archeogeography (*archéogéographie*) (Chouquer 2003b). It was defined as "the discipline of the analysis of physical, social and hybridized forms, with reference to geography, agrarian morphology and urban morphology"[6] (Chouquer 2011, § 5). The new discipline notably grew up around a research group[7], resulting in the publication of a *Traité d'archéogéography* (Chouquer 2008b; Chouquer and Watteaux 2013) and a handbook (Robert 2011). Archeogeography is now taught in several French universities, and at the *École des hautes études en sciences sociales* (EHESS)[8]. Since 1999, a dozen theses have been published in the domain of archeogeography, within connected disciplines including history, art history and archeology.

4 ...*discipline "d'interface" entre les sciences de la nature et les sciences sociales.*

5 ...*système interactif entre deux ensembles constitués par un (ou des) sociosystème(s) et un (ou des) écosystème(s) naturel(s) et/ou artificialisé(s) s'inscrivant dans un espace géographique donné et évoluant dans le temps [...] Cette notion vise à "étudier les interactions sociétés-natures par une approche intégrée et systémique à long terme, et dans une perspective de développement durable".*

6 ...*discipline de l'analyse des formes physiques, sociales et hybridées, dont les disciplines référentes sont la géographie, la morphologie agraire et la morphologie urbaine.*

7 GDR 2137 *TESORA* (*Traitement de l'espace des sociétés rurales anciennes*), CNRS, 1999-2007, coordinated by G. Chouquer (UMR 7041-Nanterre).

8 The first chair of archeogeography at the EHESS was created in 2012, with the author's own recruitment to the post, entitled *Fabrique, transmission et résilience des formes du paysage : archéogéography* (Fabric, Transmission and Resilience in Landscape Forms: archeogeography).

Following the approach laid out by G. Bertrand, archeogeographers promote a clearer articulation of both the natural and social spheres in landscape studies, based on a systematic approach. In 1997, Philippe Leveau, in *Les Formes du paysage*, created a formalization of landscape elements in the form of a spatial system, based on a set of interactions between different sub-systems: population and settlement dynamics, political systems, technical systems, exchange systems, agrarian systems, bioclimate and topography/relief. "Nature" is not taken for granted, but is seen as a construct at the meeting-point of the natural and the social. Interrelations between sub-systems of natural origin and social sub-systems follow a retroactive loop mechanism. For example, the topography/relief sub-system may be modified by that of landscape forms and by agrarian systems (Leveau 1997). In the same book, G. Chouquer suggested that the forms of spatialization studied by archeogeologists could be grouped into a morphosystem covering the whole area of planimetric landscape forms, "whether of organic or anthropic origin"[9], in interaction with both the ecosystem and the sociosystem (Chouquer 1997a, p. 18)[10] (Figure 3.1). From the 1990s on, archeogeography attempted to re-establish connections between the natural and the social, seen as separate elements in morphological analysis approaches borrowed from art history (consider, for example, the way in which P. Lavedan recommended "ignoring" geographic sites, discussed in section 1.3.4). A further aim of archeogeography is to move beyond regressive analysis and semiological analysis of landscapes, which do not allow clear comparisons of morphological results and field observations.

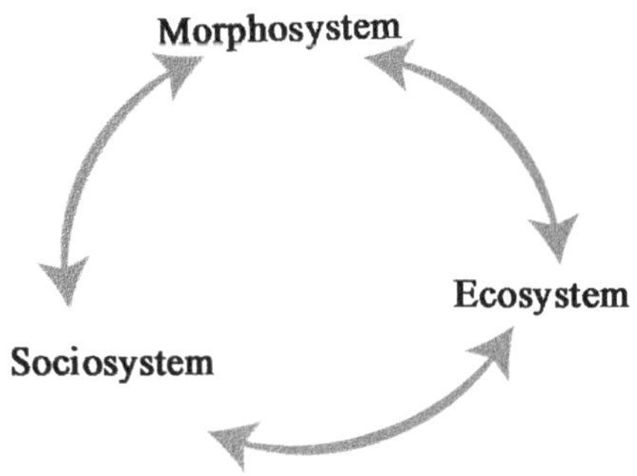


**Figure 3.1.** *The systemic approach in archeogeography as promoted by Chouquer (1997a, p. 18). The morphosystem – the object of study in archeogeography – is located at the meeting point of ecosystem and sociosystem (Robert 2019). For a color version of this figure, see www.iste.co.uk/robert/resilience.zip*

9 *...qu'elles soient d'origine organique ou anthropique.*

10 In 2003, the archeologist C. Raynaud described an articulation of the agrarian landscape with other éléments such as systems of production, exploitation, the exploited ecosystem, rural habitat, land ownership structures, etc., in a detailed model of agrarian systems (Raynaud 2003).

## 3.2. The inversion of time

### 3.2.1. *Non-linear temporalities*

The increasingly systematic comparison of field data and morphological reconstructions appears, at first glance, to offer a golden opportunity for "checking" the materialization and dating of ancient planned plot layouts, painstakingly reconstructed from the 19th century onward, using planimetric documents. In many cases, the absolute dating elements discovered in archeological structures (ditches, roads, etc.) fail to correspond to the relative dating schemes established by morphologists on the basis of map analysis and aerial photographs. For example, in the course of an archeological investigation prior to the construction of the TGV Méditerranée high-speed train line, archeogeographers were surprised to discover that one of the major axes of a centuriation, shown as a path on cadastral plans of the Napoleonic era, yielded no artifacts from Antiquity. In other places, boundaries corresponding to the regular forms of ancient centuriations appear to date from the medieval or even modern period (Jung 1999; Chouquer 2008a, p. 65).

Incoherencies between the dates assigned to landscape elements by morphologists and the dates obtained from artifacts on the ground are also encountered in the case of open fields of presumed medieval origin (section 2.2.4). Strips of land, presumed to correspond to a later division of open fields, offer up pre-medieval dates, or are traversed by Roman roads, turning the usual logic of stratigraphic interpretation on its head (Robert 2013) (Figure 3.2).

Other temporal anomalies are found when digging, for example, hiatuses in the material manifestation of ground footprints. Many cases seem to indicate that forms are not, in fact, transmitted via a linear temporal continuum. In the context of the TGV Méditerranée investigations, for example, Jean-François Berger and Cécile Jung highlighted the case of boundary ditches that have survived from ancient times to the present day, but which were periodically buried beneath sedimentary deposits, in some cases to a remarkable extent (Berger and Jung 1996) (Figure 3.3). Numerous cases of interruptions and retrieval of forms can also be found in urban environment. In Cergy, for example, we see that a new bridge built over the Oise river, during the development of the new town of Cergy-Pontoise in the 1970s, was constructed on exactly the same site as a long-departed medieval bridge, of which no material trace or even oral tradition remained (Robert 2003b). Forms may thus re-emerge at given locations after a long break in their material manifestations, and/or in the memory of a society.

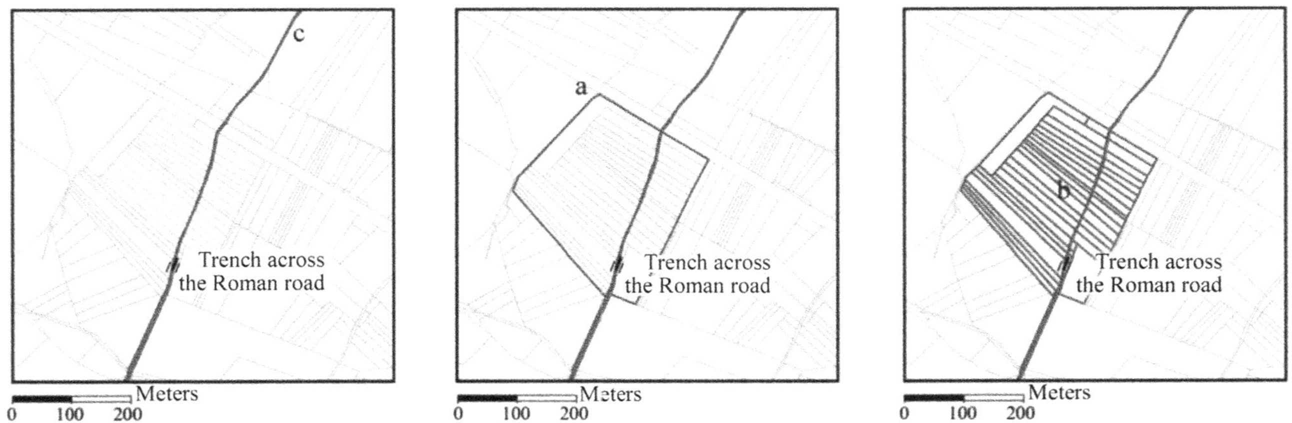


**Figure 3.2.** *In Marines (Val-d'Oise, France), the relative stratigraphy approach applied to landscape forms does not correspond with archeological findings. Road (c), which cuts across (a) and (b), is a Roman road and is thus earlier than the parcel layout seen on the cadastral plan. The "logical" relationship, according to traditional semiological analysis techniques (a < b < c) (Figure 2.8), must be reinterpreted in this case (c < a < b) (Robert 2019). For a color version of this figure, see www.iste.co.uk/robert/resilience.zip*

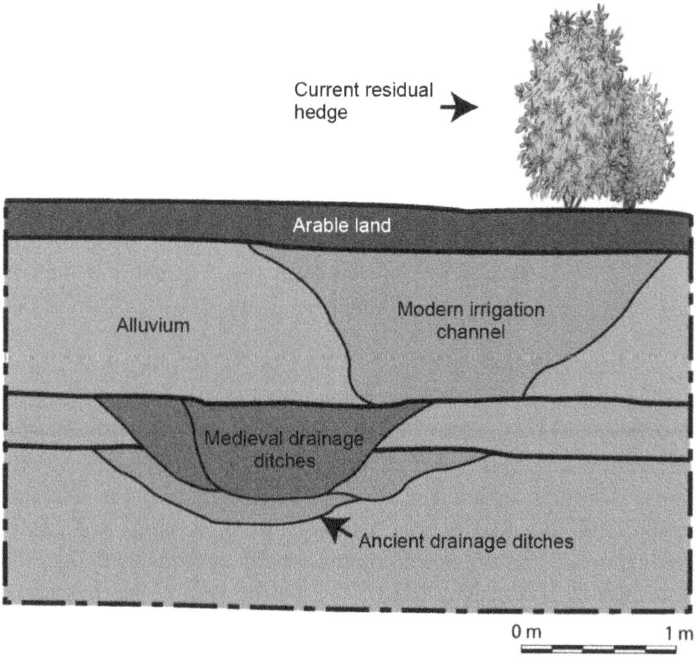


**Figure 3.3.** *In Pierrelatte, there has been a boundary on the same site from ancient times to the present day, in the form of drainage ditches, irrigation channels and a hedge, even after phases of sedimentary buildup due to a nearby stream bursting its banks (adapted and simplified from Berger and Jung (1996); Robert (2020)). For a color version of this figure, see www.iste.co.uk/robert/resilience.zip*

One of the main contributions made by archeogeography around the turn of the millennium was in showing that the persistence of forms is not solely dependent on the continued presence of visible traces on the ground, or on an uninterrupted succession of functions. A form introduced into a landscape at a time t may last and even evolve, or it may disappear completely only to re-emerge at a time when even the memory of the original form has been lost.

### 3.2.2. *Evolved forms versus decayed forms*

The reintroduction of the geographical aspects of a site played a crucial role in understanding the dynamics of the transmission of forms[11]. Within the framework of

11 The term "transmission of forms" (*transmission des formes*) progressively came to replace those of persistence or permanence of forms in archeogeography from the 1990s on. The emergence of this term will be discussed in greater detail in section 3.5.3.

the *Archaeomedes* program, for example, researchers considered the role played by ancient plot layouts near the town of Orange, France, in societal responses to flood management from the late first century BCE. They showed that developments made in the Roman era were designed to drain off excess water toward the Rhône. The form and orientation of ancient plans depended, to some extent, on soil types and on the need to regulate water flows. Some of the major axes of the centuriation were thus used to collect and drain runoff water. Researchers created a schematic map of the drainage system in the Tricastin area, overlaying ancient hydrological and ditch networks onto the orientations of the cadastral plan, combining both human and natural data into a single system directed by water flows (Van der Leeuw 1998, p. 218).

The correspondence between plot boundaries and hydrosedimentary networks led archeogeographers to consider that the centuriation patterns observed on maps of the Rhône Valley or in the Italian plains are, in fact, medieval and modern constructions as much as ancient forms. Robin Brigand, for example, showed how ancient land boundaries in Venezia were reused and developed for drainage, irrigation or land organization purposes by agrarian settlers in the medieval and modern periods (Brigand 2010). The zones where the highest density of medieval foundations is found are also those which feature the most "complete" ancient grid layouts; this observation runs contrary to the previous belief that medieval habitat was one of the main factors in the degradation of ancient systems (section 1.2.5).

Thus, forms which are visible in the present but have their roots in the ancient past cannot truly be considered as degraded or decayed states of previously "completed" forms. These are "evolved" forms: from their first origins in the past, they have undergone multiple transformations, through reuse, extension, etc. by successive societies. From the 1990s on, archeogeographers no longer attempted to reconstruct landscapes as they were in different past states (synchronous states), focusing instead on the broader trends that have structured a landscape over time. Our object of study is no longer a form at a time t – n (initial forms in a past, completed state), but rather the evolutions of this form over time (Robert 2003c).

### 3.2.3. *From regressive analysis to compiled maps*

The 1990s also saw a major transformation in the documentary sources used for morphological analysis, with the advent of "compiled" maps in archeogeography. Previously, past information was superposed onto existing maps, essentially for the purposes of localizing archeological remains[12]. In archeogeography, this form of comparison is used to study both past and present forms from a single source document. In the early 1990s, we defined the modes of construction of compiled maps (Robert 1994)[13]. The author's own thesis showed how, from a theoretical standpoint, the different temporalities seen on a map can be studied in parallel, taking account of the dynamics of forms up to the present day, in place of a regressive approach (Robert 2003b).

Different periods are shown together on a compiled map (Figure 3.4). The aim is to collate spatial and historical information for a given location, without, *a priori*, attempting to organize this information chronologically. Thus, "fossil" traces, obtained through photographic interpretation or archeological excavations, are shown alongside landscape elements found on old maps, historical references, geolocated microtoponyms, etc. The first stage is to identify structural forms based on their geometric and geographic characteristics, independently of date, function, origin (anthropogenic/natural), etc. At this stage of reasoning, only the geometric forms and precise location of objects are taken into account. This means that forms that are traditionally represented on maps with a different semiology (for example in terms of distinctions between natural or man-made elements, or functional and non-functional landscape forms) are all shown together and in the same way.

Archeogeographers work with all of the forms present on geometric planimetric documents, and not simply on presumed historical forms resulting from a highly selective triage process. Traces of "natural" origin, present in large numbers on vertical aerial photographs, are also collected. Archeological traces are considered alongside elements identified through photo-interpretation, for example, such as paleochannels that indicate that a zone was once humid[14]. The axis of simultaneities observed in this way is that of the present, but of a present enriched with all of the marks and traces of the past (Robert 2003b).

---

12 For example, archeological information was increasingly included on town plans from the 19th century on; another example is that of the English archeologists who "compiled" fossil information onto contemporary maps in the emerging discipline of landscape archeology in the 1970s (Aston and Rowley 1974, p. 173).

13 Creation of a geometric background showing old plot layouts, transfer of data from old maps and photographic interpretaiton, transfer of simplified archeological site plans on a large scale (from 1:2000 to 1:10000), then generalization to 1:250000 scale. At the time of writing, compiled maps are generally accessed in the form of geohistorical databases created using a geographic information system (GIS) (Robert 2011).

14 See, for example, the detailed examination of the former Tilles marshes near Genlis, Côte d'Or, France, carried out by G. Chouquer in the 1990s (Chouquer 1996a, p. 47).

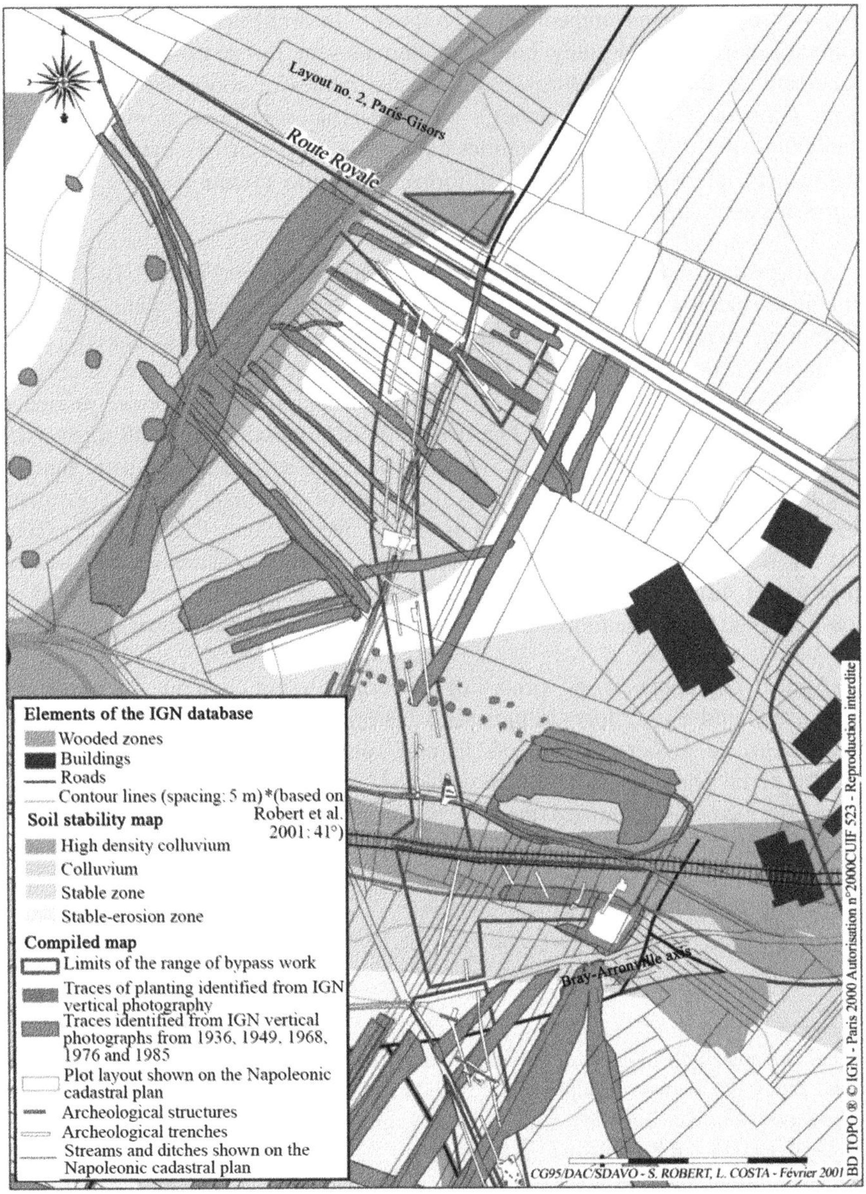


**Figure 3.4.** *Example of a compiled map of the town of Marines (Val-d'Oise). Traces identified by photographic interpretation and archeological investigations, former landscape forms shown on old maps (such as an old railroad) and the plot layout from the Napoleonic cadastral plan are all superimposed on a current map (source: Robert 2003b). For a color version of this figure, see www.iste.co.uk/robert/resilience.zip*

The resolution of observation is relatively high, as maps are compiled onto the plot layouts of the Napoleonic cadastral plan or vertical aerial photographs. The plot layout grid provides a remarkably helpful basis for recording elements. A simplified version has been developed for use with maps at a scale of up to 1:25,000 (Robert 1997). From the 2000s on, geographic information systems (GIS) were increasingly used to facilitate compilation and to allow users to easily switch between different map scales.

## 3.3. Discovering new forms

The fact of looking at objects in terms of their formal and geographic characteristics, independent of their date, function and natural/anthropogenic origins, allows us to make connections between landscape elements that are traditionally considered separately in morphological analysis. New objects emerge, notably fluvioparcellar networks (*reseaux fluvioparcellaires*) and formation networks (*reseaux de formation*), which raise questions about the relationship between forms and societies. In archeogeography, the notion of "network" was initially used in the morphological sense to refer to "an arrangement of horizontal and vertical lines" (Concise OED, 2008), later evolving toward a notion of the network as a system (discussed in greater detail in sections 3.4.2 and 3.5).

### 3.3.1. *River corridors*

Archeogeographers first began to study networks comprising both hydrographic forms and anthropogenic elements in the 1990s. Ricardo Gonzalez Villaescusa studied irrigation lines in Spain, many of which were developed at the time of the Muslim Conquest. He used a notion of "organic regularity" (*régularité organique*) to describe networks of canals which were "regular, sinuous, and constantly follow the microrelief, adapting to contour lines, to bring water to the plots they serve"[15]. These contours exhibit a certain parallelism while adapting to the relief of the land, resulting in regular metric patterns (Gonzalez Villaescusa 1997, p. 159). In the early 2000s, Caroline Pinoteau mapped forms in the west of France including active streams, paleochannels, hydromorphic zones and drainage ditches indicated on both 19th-century cadastral plans aerial photographs, where it is difficult to distinguish between natural and anthropogenic forms. Unlike topographic mapping methods, which use different conventions to represent objects according to their origins (e.g. natural hydrography in blue, human-made ditches in black), all hydrographic forms

15 *...réguliers, sinueux, et suivant constamment le microrelief en s'adaptant aux courbes de niveau, pour apporter l'eau aux parcelles qui sont desservies.*

are shown in the same way, and the author refers to the resulting "mixed physical and social object[s]" (*objet mixte physique et social*) as a "hydro-parcellar network" (*réseau hydro-parcellaire*) or as "fluvio-parcellar corridors" (*corridors fluvio-parcellaires*) in the case of corridors, which follow certain active streams or paleochannels[16] (Pinoteau 2003). Magali Watteaux developed the notion of a "hydro-vegetalo-parcellar network" (*réseau hydro-végétalo-parcellaire*) to characterize the forms which she studied in the bocage of the Vendée, France (Watteaux 2009, p. 372). Nevertheless, in current usage, the term "river corridor" has come to dominate. This notion was used by the historian Hélène Noizet to describe hybrid objects transmitted by the form of watercourses, the network of streets or paths and the plot layout (Noizet 2005, p. 116). The urban forms of both Tours and Paris, for example, bear the marks of surface water management techniques. The locations of paleochannels of the Loire and the Seine, respectively, have played host to drainage ditches, sewers, defensive ditches, then urban streets, all following the curved or sinuous form of the original channels. These forms follow the shape of the floodplain or the steepest slopes in order to optimize surface water evacuation. A comparison of morphological analysis with textual evidence shows that the forms of ancient meanders were transmitted by means of multiple mechanisms – decisions and social actions – with multiple aims: circulation, enhancement of dwellings or other buildings, financial speculation, etc. (Noizet 2005; Noizet *et al.* 2013).

In Paris, the curved form of a series of streets on the Right Bank (rue de Provence, rue Richer, rue des Petites Écuries, rue du Château d'eau, rue Amelot, etc.) (Figures 3.5 and 3.8) reflects the location of a paleomeander in the Seine, succeeded by a humid zone which was then progressively drained. The central collection channel of the medieval drainage network, the modern sewer system and the alignment of the streets above this system have reproduced the form of the paleomeander over the course of history. This persistence is the result of multiple transformations, brought about by multiple actors, from individuals up to and including the municipality. In 2006, 20.98 km of Parisian streets (or sections of streets) were identified as having forms directly influenced by that of the paleomeander. The fluvial form altered over time as a

16 The term "corridor", in this context, is similar to the notion used in ecology to define "linear elements in a landscape, of which the physionomy differs to that of the surrounding environment. They play several roles, acting as conduits which facilitate movement, or as barriers, which limit it" (*des éléments linéaires du paysage dont la physionomie diffère de l'environnement adjacent. Ils ont plusieurs rôles dont celui de conduit qui favorise le mouvement, ou de barrière qui le limite*) (Burel and Baudry, p. 351). In landscape ecology, their form is considered to be partially dependent on human action (Boissinot and Brochier 1997, p. 37).

result of major floods when the Seine burst its banks[17] and of human activity to make the humid zone more inhabitable. The observed form is a result of both the natural and social histories of the site (Noizet *et al.* 2013). The identification of river corridors in archeogeography shows a capacity to take account of objects with complex temporal origins into analyses, and removes certain traditional distinctions made between natural and anthropogenic objects.

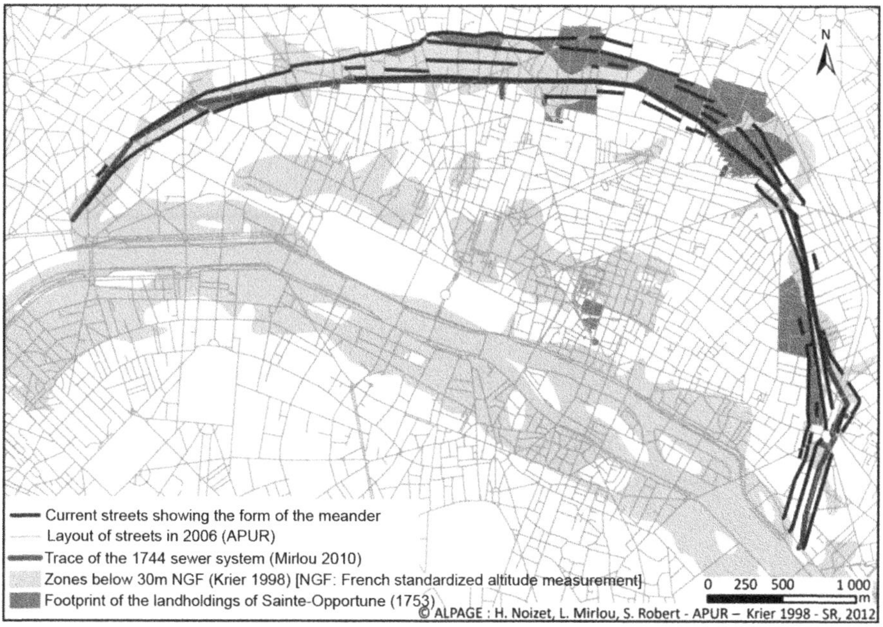


**Figure 3.5.** *River corridor on the site of a paleomeander in the river Seine on the Right Bank in Paris, visible in the form of present-day streets (Robert 2012b, using data from the ANR Alpage program, APUR, L. Mirlou, H. Noizet, S. Robert). For a color version of this figure, see www.iste.co.uk/robert/resilience.zip*

### 3.3.2. *Formation networks*

A new type of network was identified in the 1990s based on methodological developments within the field of archeogeography. Large-scale grid plot layouts were observed in different regions across France (for precise locations, see Robert 2011, p. 178). These layouts are characterized by their relatively large size, up to 200 km$^2$, and are structured by lineaments of several kilometers in length in the form of roads

17 Maps showing the extent of the Seine flood of 1910, for example, show that the hydrographic system made a partial return to its original space, despite the absence of visible traces on the surface (Noizet *et al.* 2011).

and paths, ditches, talus, etc. While relatively square in shape, the plot layouts do not follow the same strict geometric characteristics found in ancient or medieval planned plot systems. There is a certain level of flexibility in their structure, as they include local oro-hydrographic force lines (Figure 3.6). Archeological data collected during excavations of certain elements in these formation networks suggest dates from the late Iron Age to the beginning of Classical Antiquity.

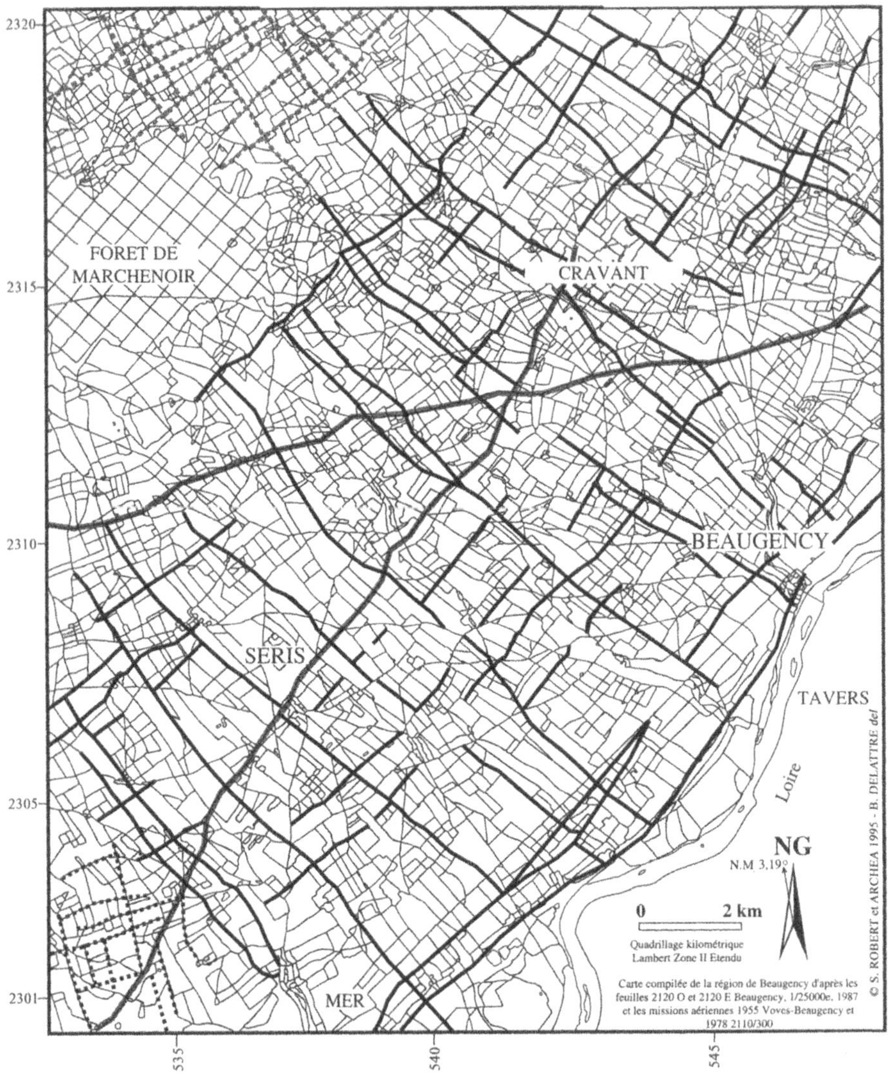


**Figure 3.6.** *Example of a formation network around the town of Beaugency, France (source: Robert 1996b)*

In the absence of written descriptions, the naming of these systems was problematic; one early suggestion was *parcellaires romano-indigènes* (Romano-Indigenous plot layouts), inspired by studies of similar plot layouts of the same period in England, the Netherlands and Denmark (Chouquer 1994). The matter was a subject of some debate in the archeological community; excavated plot layouts from the pre-Roman (Gallic) period had previously been considered as marking the outlines of limited cultivated areas, close to dwellings, which only concerned "a small community, often, doubtless, a single owner". These "islands" were believed to be separated from neighboring communities being "relatively or completely unstructured" (Mennessier-Jouannet and Buchsenschutz 1996, p. 176)[18]. The plot networks on a scale of several hectares observed on maps and aerial photographs did not correspond to this island-type organization. A distinction needed to be made between buried traces, situated around habitats identified through excavation, and the broad lines identified through planimetric documentation. These latter traces were believed to have their origins in the pre-Roman period, but to carry the marks of multiple transformations, additions and developments over the course of later periods.

A second option was to characterize these systems according to the perceived formation process, which differs from that of the planned layouts traditionally studied in morphology. Cédric Lavigne proposed that a distinction should be made between "formed" plot layouts – *parcellaires de formation* – and the "founded" plot layouts – *parcellaires de fondation* – of the ancient and medieval periods, organized "according to a norm, giving material form to a thought, will or ability on the ground"[19] (Lavigne 1997, p. 150). Dating elements collected from different sites gave dates from the Gallic period to the present day. Some of the lines in question still have a major impact on landscape organization, establishing the dominant orientations of the plot layout or road and path networks. The vast grid network around Beaugency, first identified in 1972, is still clearly visible by plane, while in Sénart, some of the developments in the new town follow past and inherited grid patterns (Liger 1972; Robert 1996a, 1996b).

These hesitations in the characterization of new discoveries highlight the difficulty of transitioning from approaches marked by classic morphology to a more dynamic vision of the processes by which forms are created and transmitted. In archeogeography, the discovery of this type of object reinforced a growing conviction that simple causality relationships between societies and forms are insufficient, and that a new paradigm is needed for morphological analysis.

---

18 *...qu'une petite communauté, souvent sans doute un seul propriétaire. [...] ...un* outfield *qui est moins, ou pas du tout, structuré.*

19 *...selon une norme, matérialisant au sol, une pensée, une volonté ou un pouvoir.*

## 3.4. Landscape as a self-organized system

### 3.4.1. *Self-organization and morphogenesis*

Work formalized in the 1970s under the umbrella term of complex thought – *pensée complexe* – drawing on information sciences, cybernetics, systems theory and the theory of self-organization (Morin 1977) cast new light on the genesis and long-term evolution of forms. In France, the theory of self-organization was first used in settlement archeology in the context of the European *Archaeomedes* program; this paradigm provided a shared ground for archeologists and geographers, the latter having adopted the notion a few years earlier (*Archaeomedes* 1998, p. 16; Pumain *et al.* 1989). Self-organization may be defined as the "process by which a set of elements interacting with one another produce an organization which tends to last for a certain period of time, in the absence of any prior intention to produce this organization"[20] (Lévy and Lussault 2003, p. 93).

According to the theory of self-organization, interactions between lower level entities can result in a higher level organization with certain permanence. A unified structure may thus emerge from a group of projects, even if some elements seem contradictory, due to a process of mutual adaptation. This definition may be applied to a town, for example, which may be considered as:

> a group of elementary actors (such as households, entrepreneurs, municipal authorities) who, through their interactions, produce urban forms at a higher level of observation (such as districts with distinct functions and compositions)[21] (Pumain *et al.* 1989, pp. 3–4).

A perceived order is not, therefore, planned from the outside, but emerges as a result of multiple purposes which mutually adjust to one another in "a disorder of genesis and creation"[22] (Morin 1977, p. 41).

The self-organization paradigm provided a helpful theoretical basis for the revision of morphological analysis practices within the emerging field of

---

20 ... *processus par lequel un ensemble d'éléments interagissant les uns avec les autres produisent une organisation qui tend à se maintenir sur une certaine durée, sans qu'il n'y ait eu volonté a priori de produire cette organisation.*

21 ...*un ensemble d'acteurs élémentaires (par exemple, les ménages, les entrepreneurs, les responsables municipaux) qui, en interagissant, produisent à un niveau supérieur d'observation des formes urbaines (par exemple des quartiers aux fonctions et aux compositions différenciées).*

22 ...*un désordre de genèse et de création.*

archeogeography. In 2000, Claire Marchand drew on the works of Edgar Morin, René Thom, Henri Atlan and Ilya Prigogine to explain the complex dynamics involved in the formation and conservation of plot layouts and road networks, considered in this context as self-organized systems. Morphogenesis, in the sense used by Morin, is the appearance of "a reality which is topologically, structurally and qualitatively new in space and time"[23] (Morin 1977, p. 115); the term is used in archeogeography to describe the conditions of emergence of both structured and structuring forms. Morphogenesis had previously been used in the 1980s to describe the emergence of structured and durable plot and road networks; however, it only took on a truly systemic sense with the work of C. Marchand in the early 2000s (Marchand 2000, 2003). The author used René Thom's catastrophe theory (Thom 1973) to describe the process leading to the appearance of organized plot networks:

> A change in form can be interpreted according to a dual principle of differentiation/integration: differentiation, as it raises forms which have a certain affinity to one another out of an anonymous morphological "magma". [...] It is also a process of integration, since the network is an organization which transforms a discontinuous variety of elements into an overall form. It distinguishes individual elements according to norms which are proper to it[24]. (Marchand 2000, p. 201)

Thus, in the early 2000s, the term "network" was increasingly used in the organizational or systemic sense within the context of archeogeography.

### 3.4.2. *Plot layouts and road and path networks as systems*

The geographer Denise Pumain defines the system as "an entity which is autonomous with respect to its environment, organized according to a stable structure (which can be identified over a certain duration), made up of interdependent elements of which the interactions contribute to maintaining and

---

23 *...une réalité topologiquement, structurellement, qualitativement nouvelle dans l'espace et le temps.*

24 *Un changement de forme peut être interprété selon un principe dual de différenciation/intégration : un processus de différenciation d'une part puisqu'il fait émerger de l'anonymat des formes qui ont entre elles certaines affinités, qui les différencient donc d'un "magma", du fond, morphologique. [...] Un processus d'intégration également puisque le réseau est une organisation qui transforme une diversité discontinue d'éléments en une forme globale. Il distingue des éléments singuliers selon des normes qui sont les siennes.*

developing the structure of the system"[25] (Pumain 2004). Foundation networks, formation networks, organic networks, river corridors and road and path networks, in the sense used in studies from the 1990s on, can all be defined as networks. They are differentiated from their environment by their morphological characteristics and have a certain durability over time, as they can be reproduced by means of interactions within the system. In plot layout networks and road networks, seen as systems, individuals interact, according to social relations of production or cultural relations, with their geographical substrate; this latter element consists of a set of hybrid forms associating natural and social forms. The new properties that emerge include the recurrence of certain morphological characteristics (orientation, geographical location, alignment, etc., described in section 3.5.1). As in any system, the properties of any new element are dependent on the others and on the system itself via a retroactive process, a "return effect from the whole toward the constituent parts"[26] (Lévêque *et al.* 2003, p. 116).

The landscape elements that make up plot layout and road network systems (roads and paths, boundaries, ditches, etc.) reproduce identical morphological and spatial characteristics by which the network may be distinguished from its environment. These characteristics tend to influence new objects arriving in, or in the vicinity of, the space covered by the network. The new extent of the network may be defined by observing the boundaries within which these morphological characteristics are repeated; these spaces may extend from a few hectares up to vast areas. For instance, foundation or formation networks are studied on a microregional scale, while road networks cover much larger areas; other morphological units may be limited to a few hectares in size (Robert 2011).

Several phenomena contribute to the reproduction of plot layouts and road networks over time. Numerous examples from both rural and urban contexts show successions of landscape elements on the same site, following the ground footprint of a boundary, a thoroughfare or the site of an activity. The model is transformed, but the ground footprint remains. The cross-section from Pierrelatte shown in Figure 3.3, studied by J.F. Berger and C. Jung, is a striking example: an ancient ditch, a medieval ditch, a modern irrigation channel and a present-day hedge were all situated on the same site. Researchers have shown that the reproduction of this boundary in the same location is due to the fact that the whole network (beginning

---

25 ...*entité autonome par rapport à son environnement, organisée en structure stable (repérable dans la durée), constituée d'éléments interdépendants, dont les interactions contribuent à maintenir la structure du système et à la faire évoluer.*

26 ... *effet en retour du tout vers les parties.*

with the implementation of a planned plot layout in the Roman period) has persisted over time (Berger and Jung 1996). Thus, the fact that certain ancient plot layouts and road networks are still visible on present-day maps and aerial photograph is explained by their transmission through new, active landscape elements.

This reproduction is "vertical" (with forms succeeding one another in the same location), but also "lateral". Archeogeographers have shown that plot layout and road networks may expand over time beyond the initial area of implantation. In Romagna and Venezia, Italy, for example, grid layouts of ancient origin extend into spaces where there are sedimentary deposits several meters above the ancient ground level, indicating that the network cannot have simply grown up "vertically". G. Chouquer proposes a distinction between "vertical transmission" phenomena, above original forms (such as the succession seen in Pierrelatte, Figure 3.3) and "lateral transmission", where an organization can spread across a space larger than that which it occupied in ancient times (Chouquer 2009, § 17–19) (Figure 3.7).

In the case of river corridors, the organization may extend beyond the talweg of the watercourse at the heart of the corridor. For example, in Paris (section 3.3.1), the morphological characteristics resulting from the past presence of a meander in the Seine extend well beyond the space of the river bed: the orientation transmitted by its form is also seen in streets and plot boundaries beyond the line of the original channel and even beyond the floodplain (Figure 3.7).

The resilience[27] of the paleomeander is not simply a vertical phenomenon, i.e. a series of objects directly above the ancient river bed. Of the 20.98 km of present-day streets which bear the morphological characteristics of the earlier humid zone (section 3.3.1), only 17% (3.5 km) are constructed directly over the main sewer (rue Richer, rue de Provence, rue des Petites Écuries, rue du Château d'eau and rue Amelot). Note that 41% (8.75 km) of the streets that follow the semi-circular form of the meander appeared after 1744, at a time when the main sewer was becoming less visible at street level; 3.42 km (16 %) are streets that still follow the morphological characteristics of the paleomeander, but were constructed after 1823, at which point the main sewer was completely invisible at street level. Other sections are extensions of existing streets, or follow orientations that are identical or perpendicular to the old sewer (Figure 3.8). The resilience of the paleomeander form in the present street layout is explained by the fact that the river corridor generated by the meander continued to develop both vertically and laterally, long after the last traces of even a drainage channel had disappeared from the surface.

---

27 The use of the term "resilience", rather than "transmission", in this case follows the usage established in our study. The use of these terms in archeogeography will be discussed in section 5.1.4.

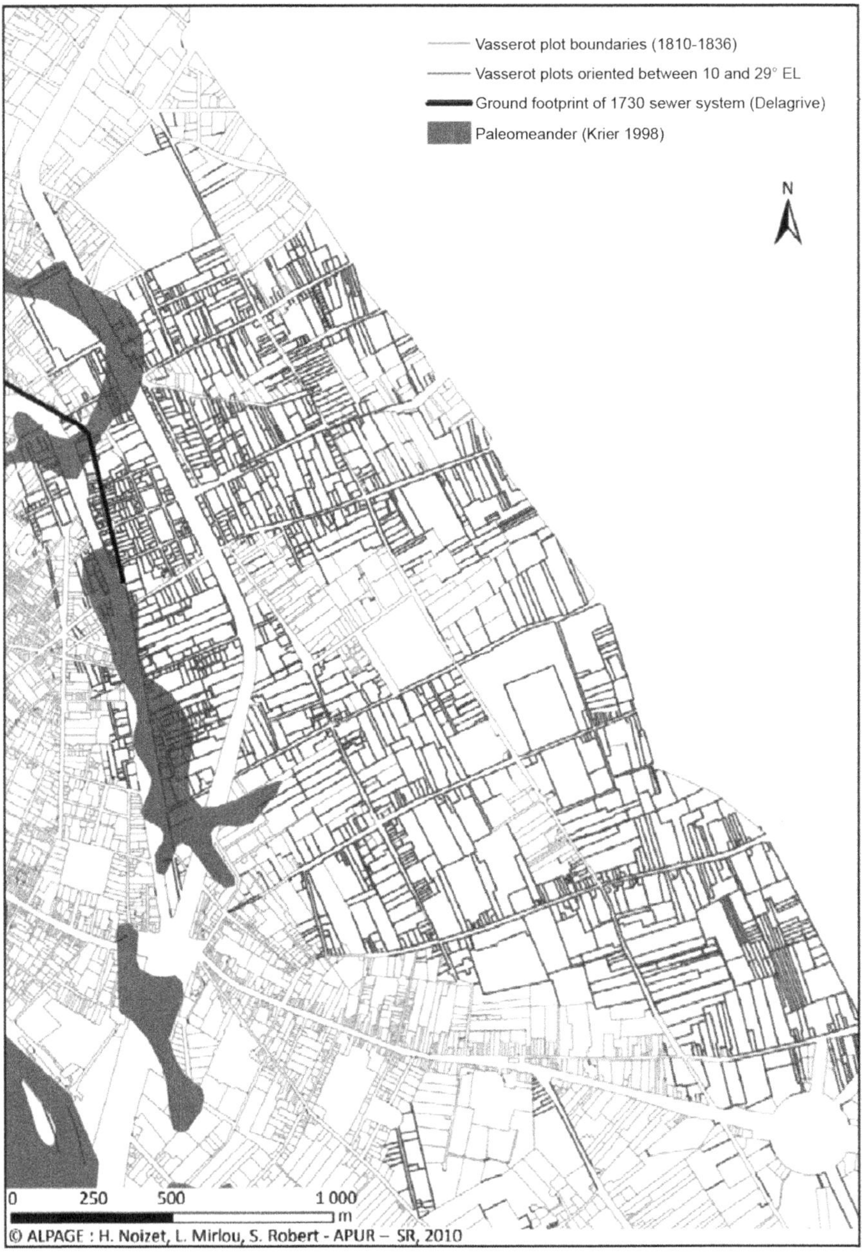


**Figure 3.7.** *Influence of the bed of a former meander of the Seine on the orientation of 19th-century plot layouts in Paris (Robert 2010; data collected from the ANR Alpage program, A.L. Bethe, L. Mirlou, H. Noizet, S. Robert)*

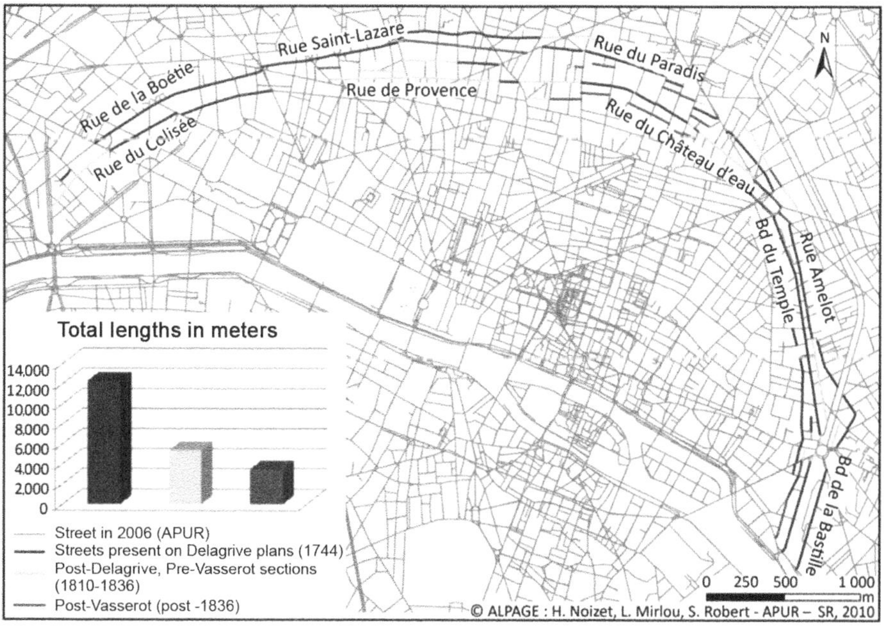


**Figure 3.8.** *Resilience of the river corridor of a paleomeander of the Seine on the Right Bank in Paris, seen in the ground footprint of present-day streets (Robert 2019; data collected from the ANR Alpage program: APUR, S. Robert). For a color version of this figure, see www.iste.co.uk/robert/resilience.zip*

Plot and road networks can also be interpreted as systems, as they show signs of reproduction in a space according to non-linear temporalities. This systemic approach provides an explanation for the hiatuses in the succession of forms, a question that had puzzled archeologists and archeogeographers ever since these gaps were first noted on the ground in the 1990s (section 3.2.1). The reproduction of the whole does not rely on the reproduction of each element in the same place and following an uninterrupted temporal continuum, but rather on the continued existence of the relationships that connect elements in a landscape. These relationships are non-linear: a key characteristic of self-organized systems[28].

28 See Pumain (2004): "In all cases, the systems under consideration are open, 'far from the point of equilibrium', and the interactions between éléments are non-linear in form". (*Dans tous les cas, les systèmes considérés sont ouverts, "loin de l'équilibre", et les interactions entre les éléments sont de forme non linéaire.*)

## 3.5. Organizing principles in plot layout and road networks

The explanatory model used in archeogeography is not, therefore, based on a human production of forms that organize a landscape from an early date and have the potential to last (planned forms). The preferred option involves an articulation of forms of both natural and social origins that form a system and generate morphological regularities, which are stable over time. This vision has led to a focus on identifying morphological principles, defined by empirical analysis, based on the observation of geographic and geometric recurrences. GIS have been used to automate this task in recent years (Robert 2011; Grosso *et al.* 2013; Robert *et al.* 2013).

### 3.5.1. *Morphological principles: reproduction of parcel layouts and road networks*

1) The recurrence of orientations[29] is one of the most striking features of the organization and maintenance of landscape forms over time. G. Chouquer coined the terms *isoclinaison* and *isoclinie* – both of which are translated into English as "isoclination" – from the adjective "isocline"[30] in order to describe the principle by which new landscape elements take on an older, pre-existing orientation:

> This is a far-reaching morphogenetic effect which finds its principal expression in the concept of isoclination: the ancient structure evolves, and other new boundaries are created, but the original orientation remains imprinted upon the plot layout, even over vast areas of land; new limits may be said to be isocline to, i.e. to possess the same orientation as, this original[31]. (Chouquer 1996c, p. 201)

29 Defined, in this sense, as the determination of the position of a location with respect to the cardinal points or any other reference (*détermination de la position d'un lieu par rapport aux points cardinaux ou à tout autre repère*) (www.cnrtl.fr 2019). For the reference systems used in archeogeography, see Robert (2011, pp. 149–152).

30 Isocline: of equal magnetic inclination (*d'égale inclinaison magnétique*) (Rey and Rey-Debove 1992, p. 1035).

31 *C'est un effet morphogénétique de très longue portée, dont l'expression principale est le concept d'**isoclinaison** : la structure antique évolue et d'autres limites nouvelles sont créées mais l'orientation d'origine reste imprimée dans les parcellaires, y compris sur de vastes superficies, et les nouvelles limites peuvent être dites isoclines, c'est-à-dire de même orientation.*

This notion applies to all types of landscape elements, from plot boundaries to buildings, roads and paths, hedges, ditches, hydrographic elements, etc.

2) The morphological principle of **isotopy** characterizes the maintenance of landscape elements in the same location, that is, at the same geographic coordinates or with "insignificant" shifts (*sans réelle portée*; Chouquer 2000, p. 147). For instance, it can be used to describe the reuse of a plot boundary, a road or dwelling structure, where the new form replaces the previous form according to a vertical succession (for example in the case recorded at Pierrelatte, Figure 3.3, or in streets built over the main sewer in Paris, Figure 3.9). Archeological examples have shown that the reproduction of dwellings or places of worship (e.g. construction of a church over an ancient *fanum*) is also a common phenomenon.

3) The principle of **iso-axiality** describes the way in which certain objects, even from very different periods, tend to align in the same direction to create a continuous line or strip (Marchand 2003, p. 100; Chouquer 2007a, p. 259). For example, a series of landscape elements with different functions and statuses may combine into a continuous form. Iso-axiality can be applied to the study of roads and paths, plot layouts, river corridors, etc. In the case of the Seine paleomeander in Paris, discussed earlier, this phenomenon is illustrated by the creation of streets over the main sewer line in the second half of the 18th century. Street sections, generally created in a relatively heterogeneous manner as part of private developments, were made uniform and brought into alignment at the request of the Parisian authorities. For example, rue Richer, created on the site of the former eight-foot wide ruelle de l'Egout, was widened to 30 feet and its alignment was modified to correspond to the end of the rue des Petites Écuries, in accordance with letters patent issued in 1784. The form of the paleomeander was maintained by isotopy (construction of streets over the lane following the sewer line) and by iso-axiality (alignment of streets to create continuity). The result is a series of streets that form a characteristic semi-circular line, marking the trace of the old Seine river bed in the very fabric of the city (Noizet *et al.* 2013) (Figure 3.9).

According to the iso-axiality principle, boundaries may also be constructed as a continuation of an existing alignment, extending the form into an area where it was not previously found. In the 18th century, for example, the avenue leading up to the Château de Cramayel was created in alignment with the main street of the village of Moissy-Cramayel (Seine-et-Marne, France). In the 19th century, a section of road D35 (the present-day D619) was created, *ex nihilo*, as an extension of the avenue leading to the Château de Cramayel (3 km)[32]. Between Moissy-Cramayel and Guignes, the present-day D619, which appears to be a continuous line, is actually the result of an iso-axial assembly of a village street (likely medieval in origin), an avenue constructed in the modern era and a contemporary section of road (Figure 3.10).

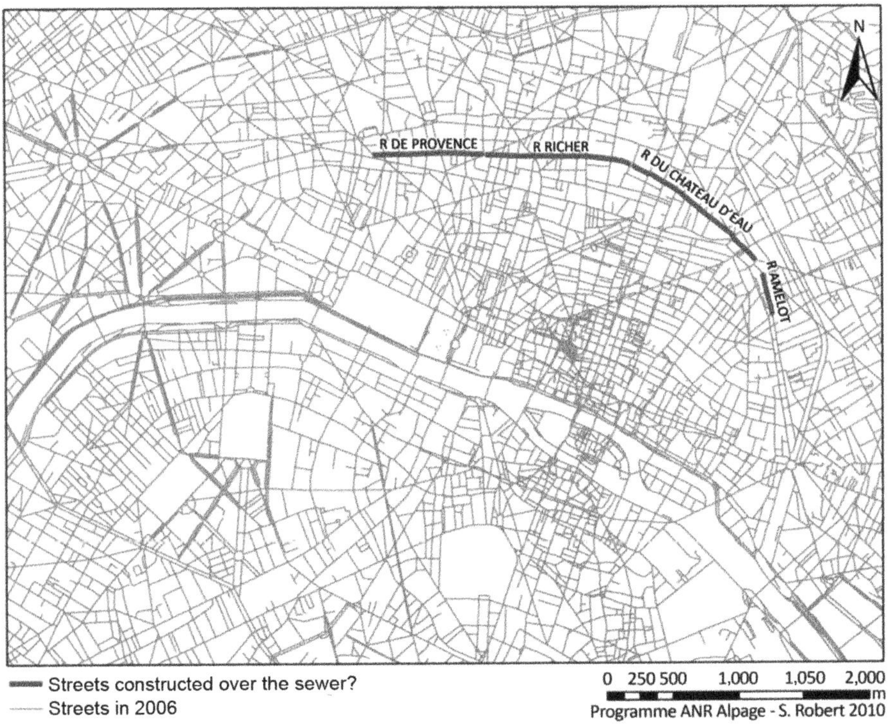


**Figure 3.9.** *Sections of streets created by private investors in the 18th and 19th centuries. By isotopy and iso-axiality, these sections reproduce the form of the location of a paleomeander in the Seine, reflected in the current street plan of Paris (Robert 2010, using data from the ANR Alpage program: APUR, S. Robert). For a color version of this figure, see www.iste.co.uk/robert/resilience.zip*

32 See *Plan d'intendance du XVIIIe siècle de Limoges-Fourches*: departmental archives of Seine-et-Marne, document 1C48/13; *Plan de la notice communale de Limoges-Fourches du XIXe siècle*: departmental archives of Seine-et-Marne, document 30Z228; and *Minutes de la carte d'état-major, début XIXe siècle,* IGN geoportal. A surveying line on the *plan d'intendance*, extending the avenue at the Château de Cramayel, indicates the planned construction of a road along this axis.

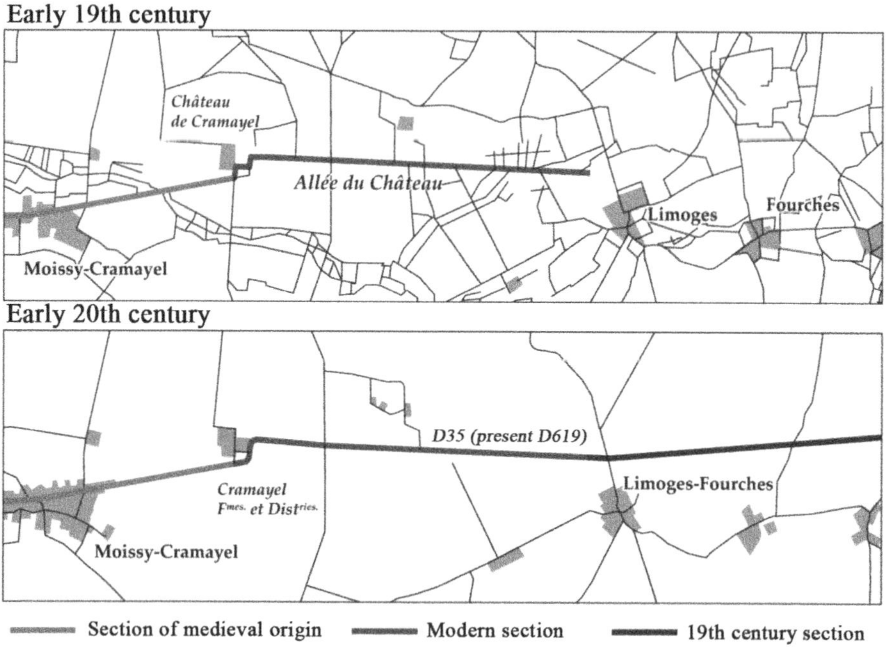


**Figure 3.10.** *Articulation and extension of road layouts of medieval, modern and contemporary origin by iso-axiality, which now make up the D619 between Moissy-Cramayel and Guignes, Seine-et-Marne, France, taken from the early 19th-century ordnance map (Minutes de la carte d'état-major) and on a topographical map from the 1950s (Robert, 2020). For a color version of this figure, see www.iste.co.uk/robert/resilience.zip*

### 3.5.2. *The role of morphogens*

The reproduction of these morphological principles in landscape elements at the microlevel results from the presence of structuring elements on a larger scale, known in archeogeography as "morphogens"[33]. A morphogen, or morphogenetic element, belongs to "the class of lineaments in a landscape which exert a persistent effort on forms, well outside their period of creation or operation"[34] (Chouquer 1997a, p. 17 and 21). This notion is used to reflect multiple observations of the role

33 The term "morphogen", coined by Alan Turing in 1952, is widely used in embryology to define elements which determine the form or structure of living organisms. In archeogeography, the notion of "spatial morphogens" (*morphogènes spatiaux*) has been used since the 1990s (Chouquer 1990, vol. I, p. 47).

34 *...la classe des linéaments paysagers qui exercent une influence persistante sur les formes, bien au-delà de leur époque de création et de leur fonctionnement.*

of lineaments in plot layout structures, where plots are created along these lines and follow the same orientation.

Several studies have shown that **ancient long-distance paths and roads** *(voies de grands parcours)* play a particularly strong morphogenetic role in the organization of formation networks (Robert 1996a, 1996b; Leturcq 1997; Marchand 1997, 2000; Robert 2003a; Chouquer 2009; Watteaux 2009). The densification and stabilization of land occupation and of transit networks, with the intensification of commercial and monetary exchanges in the period from the late Iron Age up to the 1st century AD, seem to have played a crucial role in the structuring of road and path networks and of plot layouts (Mennessier-Jouannet and Buchsenschutz 1996; Chouquer 2005, 2007b).

The resurgence of morphological analysis from the 1990s onward corresponds to a renewal of academic interest in roads and pathways. Research into historical topography had previously focused on clearly identified transit structures, such as Roman roads of major strategic importance. In 1989, the Swiss archeologist Éric Vion used contemporary topographic maps to study of ancient road networks, rather than basing his analysis on a lacunary collection of road segments identified by archeological investigation. He developed a diachronic analysis of the whole of the observable network between different towns and cities in the Vaud canton, highlighting the various branches which formed routes, or itineraries, between centers over time (Vion 1989)[35]. In this way, the focus of research shifted from the roads and paths themselves onto routes, defined here as the itinerary, or the succession of locations that are passed through, between two regional centers. From the 1990s on, archeogeographers also highlighted groupings of multiple coherent roads and paths forming part of regional routes (see Robert 2009 for a summary). These directional lines had the effect of "imprinting" major orientations onto the landscape (Figure 3.11). At a local level, the relative stability of these ground footprints over time seems to have contributed to the emergence of durable plot layouts.

The *Dynamique et résilience des réseaux routiers en région Île-de-France* (*Dynarif*) program resulted in a characterization of these informal routes, obtained by compiling stratigraphic observations from the Île-de-France region (around Paris) and proposing new methods for identification and study (Robert and Verdier 2014). Not all major itineraries appear to have played a morphogenetic role, and in some

35 Vion notably drew on the work of B. Lepetit, who analyzed the post road network in the modern period (Lepetit 1984, p. 19), see also (Verdier and Bretagnolle 2005). In the medieval period, researchers talk of a *chevelu routier* – literally "hairy roads", an image intended to evoke the spidery nature of multitudes of tiny footpaths. This idea has also been conceptualized as "road areas" – *aires de route,* or *area di strada* in Italian – which describe groupings of parallel roads and paths going in the same direction (Perol 2007).

cases, only part of a route seems to have been a morphogen. In most cases, there is evidence of a complex relationship in which local topography also plays a part. In cases where the orientation of a road or path bears a relation to surface water circulation mechanisms, there is a stronger chance that these elements played a morphogenetic role with respect to the plot layout, and the forms are more likely to have persisted (Robert 1996b, 2003b) (section 3.5.4).

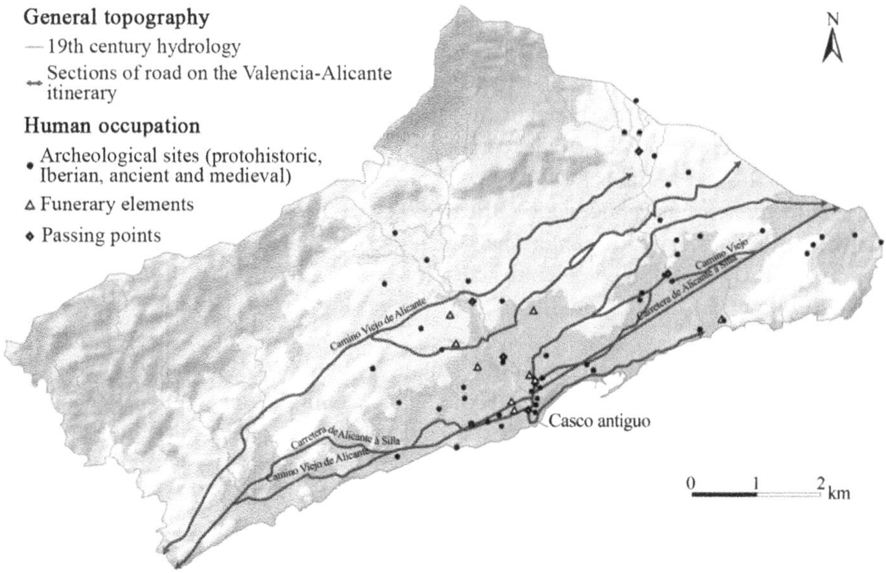


**Figure 3.11.** *Grouping of multiple road and path layouts on the route from Alicante to Valencia in the area around Villajoyosa (Spain) (L. Costa and S. Robert 2019)*[36]*. For a color version of this figure, see www.iste.co.uk/robert/resilience.zip*

The **hydrographic network** (rivers, streams, ditches, bank boundaries, etc.) is another source of morphogenetic lineaments. Hydrosedimentary flows appear to have played as important a role as road networks in the structuring of observed systems. The drainage requirements of particularly hydromorphic areas may also have accelerated the connection of formerly independent plot groups via large drainage ditches, created to improve the management of run-off. The role of surface water management or irrigation systems is clearly visible in the structure and resilience of ancient and medieval foundation networks (Jung 1999; Chouquer 2009; Brigand 2010) and in formation networks (Robert 1996a, 1996b; Chouquer 2007b; Watteaux 2009). Certain settled areas located on interfluves may have acted as "geometric morphological connectors" (*connecteurs morphologiques géométriques*)

36 For details, see Robert and Costa (2014).

between habitats and river corridors or hydroparcellar networks, by means of the vast ditch systems found in these areas (Pinoteau 2003, p. 256). This seems to have been the case for certain trench-enclosed habitats established during the 4th–1st century BCE, when hydrological developments meant that lower floodplains were becoming inhabitable and earlier channels were becoming blocked (Chouquer 2007a, pp. 337–338). Certain fortified habitats from the medieval period may also have played this role (Pinoteau 2003; Cavanna 2016). In some cases, individual plot groupings, planned or otherwise, can also be considered to play a connective role between river corridors (Pinoteau and di Pietro 2003, pp. 276–277).

The most influential element of the hydrographic network in terms of organization is not necessarily the "strongest", or even the most recent. For example, in the area around Marines (Val-d'Oise, France), the secondary network, rather than the main river, is morphogenetic (Robert 2003b). In other cases, former channels, now filled in, are more influential than the present course of a river, as we have seen in the context of the Seine paleomeander, where part of the early 19th century plot layout follows the orientation of a long-lost form (Figure 3.7).

Other elements may have a morphogenetic influence on landscape forms without necessarily constituting lineaments. Generally speaking, **slope gradients** and **major orientations** in the natural relief of the landscape (ridge lines, hills, hillsides, etc.) play a morphogenetic role, influencing the orientation of plot layouts and of the road network. Physical elements may offer possibilities for communication in a non-deterministic manner: an inherited spatial context is transformed by successive societies. As a general rule, all forced passing points, such as fords, mountain passes, bridges, city gates, etc., are also powerful morphogens as they restrict circulation (Figure 3.12). For example, funnel-shaped (or "crow-feet") road and street layouts often reflect the morphogenetic effect of a city gate, long after the wall itself has been demolished.

Certain centers of settlement also have a morphogenetic effect on road networks or plot layouts. The diffusion of plot boundary lines, associated with dwellings, through a space can result in a type of "bottom-up" structuring. There are many examples of plot layouts from the proto-historic, ancient and medieval periods that follow the orientation and, in some cases, the metrology of dwelling enclosures, radiating out into the surrounding space. The extension of these plots, "under the morphological influence of dwelling enclosures"[37] has a limited spatial extent. However, above a certain habitation density, a threshold seems to be crossed in terms of continuous connection of different elements, such as dwelling enclosures, roads or paths, and plot groupings connected to dwelling-places (Marchand 2000, pp. 204–205; Chouquer 2007a, pp. 337–338). Plots and lineaments influenced by

37 *...sous influence morphologique des enclos de l'habitat...* François Favory (Favory 2012).

dwelling enclosures eventually join up into networks (Chouquer 2007b; Favory *et al.* 2012).

**Figure 3.12.** *Morphogenetic role of topographical elements on the road network. The street layout of Argos (Greece) follows a characteristic funnel or crow-feet pattern, marked by forced passages over the river, to the north, and between the hills to the west (S. Robert 2014)*

Finally, certain morphogens result in forms which are less immediately visible, but which may have a lasting influence on landscape forms. These include **borders**, **land-holding boundaries**, **building restrictions**, etc., established by legal limitations and creating constraints that strongly influence the spatial context in which new forms are created. These may also be seen in the spatial configurations of dwellings, plot layouts, road networks, etc., which follow or are affected by these "invisible" obstacles.

Morphogens, be they of spontaneous origin or a result of human decisions (road construction, creation of a center, legal boundaries, etc.), may thus be defined as intermediary structural elements, between elementary landscape forms (sections of road, plot boundaries, ditches, etc.) and the networks that structure the landscape into an organized system. Morphogens transmit morphological principles, ensuring the resilience of plot layouts and road networks. Only some morphogenetic elements date from the emergence of plot and road systems; most other elements adopt these principles at a later date. This explains the wide variety of dating information obtained via excavations.

### 3.5.3. *Morphostasis, transformission and resilience*

In her 2000 thesis, C. Marchand used the concept of morphostasis to describe the dynamic processes at work in the formation and reproduction of plot layouts and road networks of proto-historic origin in the western Sénonais (Seine-et-Marne, France). In cybernetics, morphostasis is defined as "the processes in complex system-environment exchanges that tend to preserve or maintain a system's given form, organization or state" (François 2011, p. 396). The term was adopted in archeogeography based on the work of Edgar Morin, who defined morphostasis as the way in which an organization "maintains the permanence of the system in its form (*Gestalt*), existence and identity"[38] (Morin 1977, p. 131), and on that of Jean-Louis Le Moigne, who considered that morphostasis corresponds to phases of regulation and adaptation within a system, as a function of the permanent or changing character of its environment and of the system's purpose (Le Moigne 1994, p. 197). Marchand identified three main principles enabling the reproduction of plot layouts and road networks over time (Marchand 2000, 2003): the existence of:

– elements which play a decisive role in the conservation and diffusion of spatial information, notably by contributing to the maintenance of a certain structure on a macroscopic scale[39];

– a minimum number of elements with the capacity to preserve and transmit the spatial information contained in the morphological principles described above;

38 *...maintient la permanence du système dans sa forme (Gestalt), son existence, son identité.*

39 This is the role played by morphogens, notably those associated with circulation networks: the hydrographic network and the road network.

– a certain degree of variation in these elements, so that the same message is transmitted in different ways, resulting in "a redundancy of information, creating a reserve or stock, giving the network the means of withstanding events and maintaining its identity"[40] (Marchand 2000, p. 198)[41].

Marchand drew on the "order from noise" principle, put forward by the physicist Heinz von Foerster and popularized in France by the philosopher Henri Altan (1979), to show that transformations in plot layouts or land usage do not necessarily result in the disappearance of the network as an organized form. She hypothesized that a plot layout network will draw on "environmental disturbances as a source of regeneration and reproduction over time"[42], and that the evolution of these networks over time is based on their capacity to resist "noise", producing morphologically identical elements (the redundancy phenomenon), while also using and integrating different elements found in the "noise" (diversification phenomenon). Marchand considered redundancy as "a store of memory, critical to the integration of new elements. Without sufficient initial redundancy, the new elements will simply constitute dispersion: separate, disordered diversity"[43] (Marchand 2000, pp. 207–208). The lasting quality of a network is thus considered to depend on both its capacity to resist disturbances by means of morphostasis, and its capacity to integrate, or even generate, disturbances in order to increase its complexity through morphogenetic processes (Marchand 2000, p. 216). This echoes certain characteristics of resilience highlighted by the *Archaeomedes* program and included in archeogeography glossaries in the early 2000s, where it is defined as:

> the capacity of a set of forms (a network of lines, points, etc.) to maintain its structure despite changes in social formations and the transformations resulting from these changes. Resilience is a principle of stability in change, up to a point of rupture or division[44]. (Chouquer 2003b, p. 302, 2000, p. 190)

40 *...une redondance de l'information constituant une réserve, un stock, susceptible de permettre au réseau de résister aux événements et de maintenir son identité.*

41 For example, an orientatin which is transmitted simultaneously by a road network, plot layouts and the hydrographic network will be more resilient than an orientation transmitted by just one type of landscape element.

42 *...perturbations issues de son environnement comme source de régénération et de reproduction dans le temps.*

43 *...une réserve de mémoire, qui est indispensable à l'inscription du nouveau. Sans redondance initiale suffisante, le nouveau ne sera que dispersion, une diversité séparée, désordonnée.*

44 *...l'aptitude d'un ensemble de formes (un réseau de lignes, de points, etc.) à maintenir sa structure alors que les formations sociales ont changé et le transforment. La résilience est un principe de stabilité dans le changement jusqu'à un point de rupture ou bifurcation.*

The concepts of morphostasis and resilience are thus aligned with the notion of "transformission", a term coined by G. Chouquer (2000) to express the fact that "the system does not reproduce itself in an identical form; it lasts by self-transforming"[45] (pp. 109–110). This is supported by the fruits of cross-fertilization between the domains of morphology and archeology (section 3.1), which show that the plan of forms is transmitted *because* their shape transforms. "Transformission" is formed from "transmission" and "transformation", and is used to describe, "without having recourse to periphrases, the dual action of transformation of geographic realities in time, and the transmission of certain characteristics of these realities, giving the impression of a durability of form" (Chouquer 2003b, p. 303)[46]. Nevertheless, the term "transmission" is more widely used in geography to express notions of durability explained by change (section 5.1.4). C. Marchand used the expression "transmission of information" (*transmission de l'information*) in her thesis to describe the action of the morphological principles of isoclination, isotopy and iso-axiality (described above). These principles transmit information concerning location, orientation, alignment, etc., which appear in the course of network morphogenesis.

In the author's own thesis, from 2003, inspired by the work of C. Aschan-Leygonie, the concept of resilience was used rather than morphostasis or transformission in studying the role of transformations at different levels in the resilience of regional transit networks from Antiquity to the present day. Analyzing nine regional routes of ancient or medieval origin in the present-day Val-d'Oise *département*, France, we showed that the resilience of regional routes is dependent on the articulation of three different levels of forms, with different rhythms of transformation over time (Robert 2003b).

### 3.5.4. *The articulation of flow, ground footprint and shape*

Our study of routes in a long-term context showed that their resilience depends on the articulation of different spatial and temporal dimensions. Regional traffic regularly conflicts with local realities in the form of attempts to control parts of a route, the presence of obstacles such as city walls, topographical limitations for certain forms of transport, densification of the built landscape, insufficient road width, etc. During the Middle Ages, the courses of Roman roads were "captured" on a number of occasions as new population centers attracted increased circulation. Cases of regional circulation being captured, or drained, by towns, villages, castles or religious sites, which then profited from associated revenues in the form of taxes,

45 ...*le système ne se répète pas à l'identique ; il se pérennise en se transformant.*

46 ...*sans avoir à chaque fois à recourir à une périphrase, la double action de transformation dans le temps des réalités géographiques et de transmission de certains caractères de ces réalités donnant l'impression d'une pérennité de la forme.*

tools, markets, etc., have been identified in many regions of France[47]. However, the capturing of major road axes by local centers of population did not necessarily undermine the status of regional routes; instead, the phenomenon appears to have contributed to the resilience of the route, notably by encouraging road maintenance work in the absence of any centralized authority with responsibility in this area. Close proximity to a major route was also an important factor in maintaining local habitat, particularly around crossroads and passing places such as fords, bridges or passes.

Conflicts of form between regional traffic and the territory through which it flowed resulted in the transformation of road courses (creation of new axes, deviations, etc.) and shapes (alterations to increase serviceability, creation of new types of shape, etc.)[48]. A regional route is not limited to a single road or path object in time; it consists of multiple branches or layouts, which undergo transformations in their shape to adapt to the evolution of transportation methods and the intensity of traffic. The interactions between regional traffic and the supporting route, through a succession of layouts and shapes, create a dynamic that promotes resilience at both local and regional levels, with each acting as a contributing factor in the conservation and transformation of the other. We refer to this articulation between the three spatial dimensions of road networks as "the flow-ground footprint-shape articulation" (*l'articulation entre flux-tracé-modelé*), which could usefully be generalized to other types of forms, such as plot layout networks (Robert 2003b).

While morphologists were quick to note the shift between the transformation of the shape of landscape elements and that of their ground footprint[49], the lateral extension of networks and the re-use of forms within a non-linear temporal framework is harder to explain without using a third dimension: that of flow (section 3.4.2). Circulation, or flow, is the level of articulation between elementary entities (landscape forms) and the emerging form (the network). Flow concerns elements in a state of movement, and is characterized by "an origin, a destination and a trajectory" (*une origine, une destination et un trajet*) (Brunet *et al.* 1993, p. 219). It describes a circulation. Originally applied to liquids (e.g. the ebb and flow of the sea), the notion of flow is used in geography to denote a shift, equivalent to the "consumption" of a distance, which may be marked by a time factor (distance-time) and a cost factor (distance-cost) (Pinchemel and Pinchemel 1994, p. 98).

47 Such as the market towns which grew up around castles, *bourgs castraux* (Bur 1993). P.Y. Laffont highlights the economic role of these "captures" in the case of castles (Laffont 2008).

48 See the case of the Marines bypass in the Val-d'Oise (section 3.5.5).

49 For instance, P. Lavedan drew attention to the reconstruction of towns according to their former plan in the wake of the First World War (section 1.2.2).

Flow plays an important role in the resilience of plot layouts and road networks, as the landscape elements which make up these systems are considered as part of higher-level networks. For example, a plot boundary is not simply a point of interaction between two neighboring proprietors if it also forms part of a network through which water, people, animals and/or plants circulate. While plot layouts are generally considered as a "pavement" or patchwork (Pinchemel and Pinchemel 1994), they are also a substrate for circulation. Water and sediments flow through boundary ditches or ploughed furrows (Figure 3.13), and field boundaries were long used as circulatory routes through areas with no established pathways[50].

**Figure 3.13.** *Flooded fields in Maffliers (Val-d'Oise), January 2001. Elements in a landscape are hybrid objects, and field layouts are not simply a patchwork or tessellation covering the land. At the plot boundary, water circulates through the deepest furrow. Further on, the path along which meltwater usually runs has been temporarily transformed into a stream (photographs: S. Robert, 2001). For a color version of this figure, see www.iste.co.uk/robert/resilience.zip*

Elements in a landscape should thus be seen as hybrid objects: a road may be used for human circulation, but also for water, sediments, flora and fauna, which circulate or are transported along ditches or roadsides. A plot boundary is a limit, but also a conduit along which water, people, flora or fauna may circulate, depending on the material form of the boundary (ditch, hedge, boundary furroz, etc.) (Figure 3.13). Buildings may "cement" plot layouts (section 1.2.2), but also orient the circulation of people, air, etc. (Figure 3.14).

50 Samuel Leturcq and Blandine Vue both highlight the importance of circulation along ploughing ridges, strips of land which are created as a result of the involuntary accumulation of ploughed land along field boundaries (Vue 2002; Leturcq 2008).

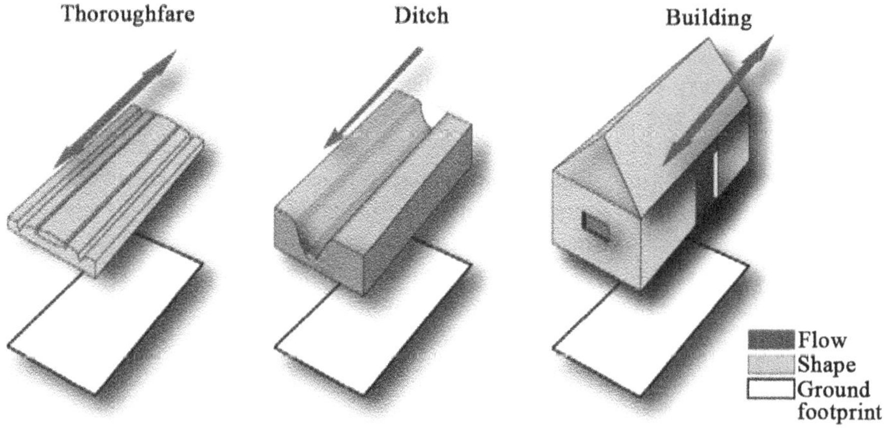


**Figure 3.14.** *The three dimensions of landscape elements: flow, ground footprint and shape. The asynchronous relationship between these three dimensions enables constant transformation at local level in response to use, while also enabling the reproduction of the morphological principles that ensure the resilience of networks at the macrolevel (Robert 2019). For a color version of this figure, see www.iste.co.uk/robert/resilience.zip*

Many roads, paths and field boundaries thus form part of "fluvio-parcellar corridors, flow spaces with complex dynamics"[51], which contribute to the resilience of self-organized networks (Pinoteau and di Pietro 2003, p. 273) (section 3.3.1). The collapse or blockage of a ditch at local level may result in malfunctions at a higher level, across a whole drainage or irrigation network. The blockage of a road or bridge may affect circulation over a much wider area. Hence, in archeogeography, elements in a landscape are not only studied in a planar (ground footprint) or volumetric (shape) framework, but also in terms of their capacity to act as substrates for flows or circulations. A network is not simply a "net", respecting the original sense of the word (section 3.3), but also a set of material elements connecting different locations across a territory, a sense it took on in the course of the 20th century: a set of "linear, permanent or temporary elements which attest to the existence of relations and guarantee the possibility of these relations, alongside nodal elements which are crucial to the organization of flows and to the operation of the system in which the network operates"[52] (Chapelon 2004, p. 2). By taking account of the "circulatory" aspect of plot layouts and road networks, it becomes possible to consider them as self-organized systems within the framework of archeogeography.

51 *…corridors fluvio-parcellaires, espaces de flux et lieux de dynamiques complexes.*

52 *...d'éléments linéaires, permanents ou temporaires, qui traduisent l'existence de relations et en garantissent la possibilité, mais également d'éléments nodaux nécessaires à l'organisation des flux et au fonctionnement du système dans lequel s'inscrit le réseau.*

The three different levels, flow, ground footprint and shape, operate as a system. The fluidity of circulation depends, to a great extent, on the form of the layouts and shapes which carry this flow. Inversely, the volume of the flow circulating along ground footprints and shapes is decisive in their transformation. Increased water circulation in a ditch or traffic on a road or path will generally lead to transformations in terms of their construction (erosion, improvement of the road, etc.) and in their ground footprint (widening, movement, addition of an additional lane, etc.). In a manner of speaking, flow, ground footprint and shape are associated in **feedback loops**, insofar as transformations of ground footprints and shapes at the microlevel have an effect on flow at the macrolevel, and vice versa (Figure 3.15). This process of reciprocal adaptation is not always synchronous over time; consider, for example, the number of roads that are insufficiently adapted to actual traffic levels.

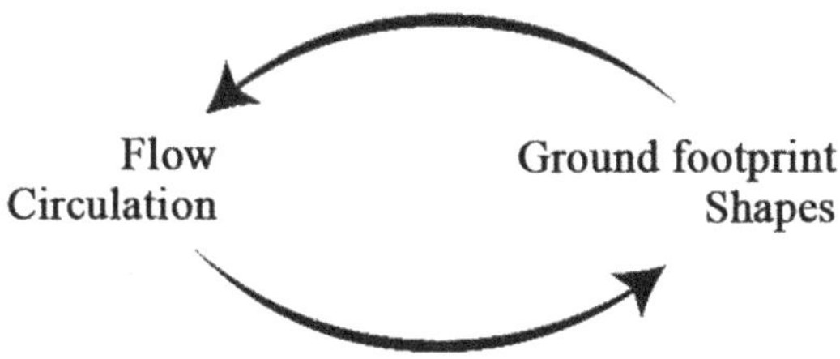


**Figure 3.15.** *Feedback loops connecting flows (circulation), ground footprints and shapes. These loops involve an adaptation of circulation to the forms which support it. This adaptation is not synchronous, and there may be significant observable time lags (Robert 2019). For a color version of this figure, see www.iste.co.uk/robert/resilience.zip*

These three dimensions correspond to different levels of observation. Classic morphological analysis tends to focus on the ground footprint level, using plot layout plans or topographic maps, while in archeogeography, ground footprints are studied within the framework of interactions with both higher and lower levels[53]. This approach has resulted in a deconstruction of certain traditional geohistorical models. For example, analyzing layouts in the southern Vendée, M. Watteaux revised the traditional notion of the bocage, considering the presence of hedges (shapes) separately from plot boundaries (layouts). While hedges may date from a variety of

53 In studying regional routes across the Val-d'Oise, for example, we used a GIS to compare the reconstruction of flows at regional level, modeled from ancient urban centers (macro scale), with sections of paths and roads from the Napoleonic cadastral plan (meso scale) and shapes obtained from stratigraphic cross-sections in archeological excavations (microscale) (Robert 2011, 2013).

periods, notably from the modern period, plot forms may be proto-historic in origin and may have originally been associated with open-field systems (Watteaux 2009). In urban environments, asynchrony in buildings, plot layouts and road and street networks has also been presented as a driver for resilience in the fabric of towns and cities (Hermenault 2017; Noizet and Clémençon 2020). The articulation of the three different levels – flow, ground footprint and shape – is possible because of the existence of different rhythms of transformation.

### 3.5.5. *Asynchrony in flows, ground footprints and shapes: the importance of lag*

A detailed analysis of regional routes shows that circulation between major centers is maintained over long periods (up to 2000 years, in six of the eight routes studied in the Val-d'Oise). The transformation of elements on lower scales – ground footprints and shapes – results in flows being maintained (Robert 2003a). At ground footprint level, a wide variety of temporalities are observed, including very long, medium or short durations and non-linear temporalities. Certain sections of Roman roads, for example, are still in use today. In the French Vexin, certain parts of the *chaussée Jules César*, constructed in the 1st century AD, have been developed and re-developed on the same site over the last 20 centuries. Looking elsewhere, the route from Paris to Rouen does not follow the course of the earlier Roman road, but features sections inherited from the medieval and modern periods (six centuries earlier in the case of the old Paris-Rouen path, two centuries in the case of the royal road, which replaced the Roman *chaussée Jules César*). Elsewhere, we find sections of road constructed much more recently (the route from Paris-Rouen via the A15 highway at Cergy, for example, is around 40 years old). Hence, ground footprints may present multiple temporalities, but also non-linear temporalities: in certain cases, a contemporary road section may "re-activate" an older, long-disused trace.

In Marines (Val-d'Oise), we analyzed the resilience of the route between Paris and Dieppe, known to have existed since at least the Middle Ages[54]. A detailed analysis, combining planimetric documentation and archeological data, showed the evolution of this route across the French Vexin (Robert 2003b). There were at least six different branches of this route in the Middle Ages, one of which dates back to the High Middle Ages (Devals 2003). The 19th-century toponymy of the roads and paths bears the marks of circulation outside of the immediate local area, for example

54 Circulation on this route is well documented, notably from the 14th century on, when *chasse-marée* – seafood transporters – used it to transport fresh seafood from the port to the capital (Robert 2018).

in the *Chemin de Gisors à Pontoise*, the Gisors-Pontoise road[55]. However, when the Paris-Dieppe road was developed in the 18th century, work focused on a ground footprint running further east, via the château of Marines. The old royal road – now a national road – retained its importance and, from the 1990s on, bypasses were created to limit traffic circulation through several towns and villages along the thoroughfare.

In the early 2000s, traffic through the small town of Marines had reached saturation point at around 12,000 vehicles/day, including trucks. Circulation was slowed by the presence of traffic lights and a variety of obstacles (roadside parking, pedestrian crossings, etc.). A bypass opened on November 12, 2002, reactivated part of the *chemin de Gisors à Pontoise*, which had survived in the form of a simple rural lane and as an administrative boundary. The old road thus regained an important function in the long-distance circulation network, following a hiatus of several centuries (Robert 2003b)[56]. While circulation at a regional level appears to follow a certain temporal continuum (over at least 700 years, in this case), the ground footprint and shape of the *chemin de Gisors à Pontoise* follow a non-linear temporality in terms of function, regaining a role in long-distance circulation following a hiatus of around 250 years (Figure 3.16).

Similarly, in the case of the bridge at Cergy, presented earlier (section 3.2.1), the heavy traffic resulting from the development of the new town led to the reactivation of an old passing-place. The setting of the new bridge was also influenced by the inherited spatial context, and other "legacy" elements were also reactivated, including a shoal in the river, facilitating the creation of a passing place, and two paths converging toward the river. The former *chemin Perryolle cy devant le chemin des Voitures* and the *sente Saint-Marguerite cy devant un Grand Chemin*[57] were redeveloped into one of the main thoroughfares of the new town of Cergy (boulevard de l'Hautil). The substrate (layout and shape) was transformed to adapt to the flow, while the increased flow led to the re-establishment of the bridge following a hiatus (Figure 3.17). In the case of ancient centuriations studied in France and Italy, the way in which plot layout boundaries played a role in water circulation is a main reason for the reuse and development of certain forms, even following periods of significant sedimentary buildup (section 3.4.2).

55 *Minutes de la carte d'état-major*, 1:10 000, 1820.

56 This example is reinterpreted using the panarchic model in section 8.6.

57 *Plan et arpentage général de la Seigneurie de Sergy près Pontoise appartenant à Messieurs les Religieux de l'Abbaye royale de Saint-Denis en France, avant 1731* (A. N. N I Seine-et-Oise 74). The toponyms indicate that these paths were in frequent use, notably for transporting goods.

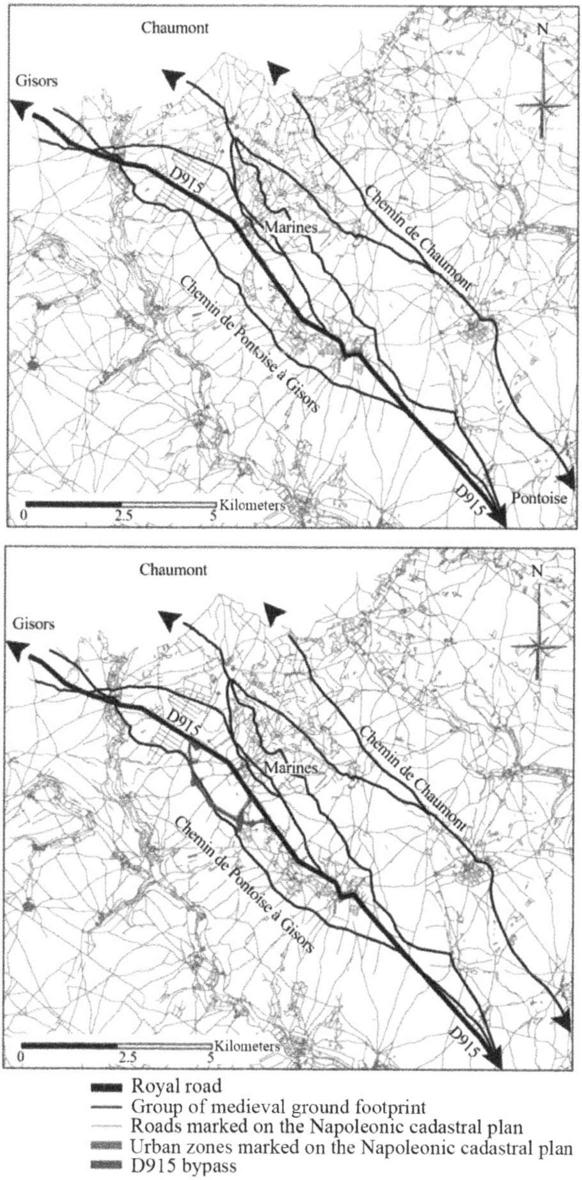


**Figure 3.16.** *In Marines, an old section of path was "resurrected" for regional circulation purposes following a hiatus. The medieval Paris-Dieppe route included up to six possible branches, including the chemin de Pontoise à Gisors. The road fell into disuse following the creation of the royal road in the 18th century, but the ground footprint was reactivated with the construction of a bypass in 2002 (S. Robert 2003). For a color version of this figure, see www.iste.co.uk/robert/resilience.zip*

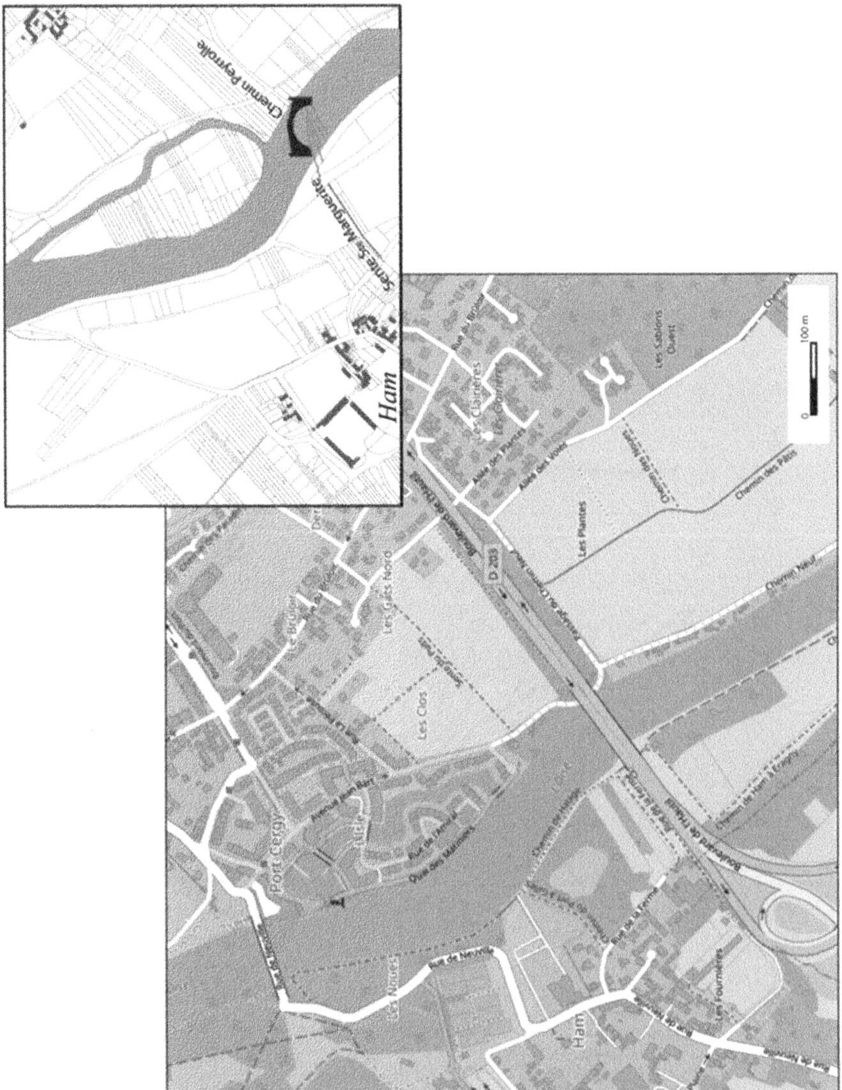


**Figure 3.17.** *Cergy: development of a new bridge on the site of a long-forgotten medieval bridge. The construction of the new town reactivated circulation on the Saint-Martin plateau, and designers "accidentally" revived the former passing place (Robert 2019; present-day map: Open Street Map). For a color version of this figure, see www.iste.co.uk/robert/resilience.zip*

Flow (circulation) plays a decisive role in the resilience of plot layouts and road networks, as it is directly connected to the morphological principles that enable reproduction of the network over time. Thus, the circulation of water down a slope, the circulation of individuals in a given direction, etc. determine, and tend to reproduce, the dominant orientations in a landscape (isoclination principle); they also tend to remain at locations which contribute to the interconnection of elements (isotopy). Similarly, the need to connect locations promotes the alignment of elements in a landscape and the extension of alignments according to the principle of iso-axiality (as seen in the case of streets constructed over the main sewer in Paris and later aligned, or the D35 road between Moissy-Cramayel and Guignes described above).

Flow enables the synchronization of elements with very different temporal origins into a network that operates in the present. For example, current regional traffic (flow) between Paris and Rouen via Pontoise travels along sections of the 1st-century *chaussée Jules César*, sections of medieval and modern roads, the A15 highway, etc. Regional circulation is possible because of a certain level of uniformization of this composite road, creating a coherent axis despite the varying origins of different sections (Figure 3.18).

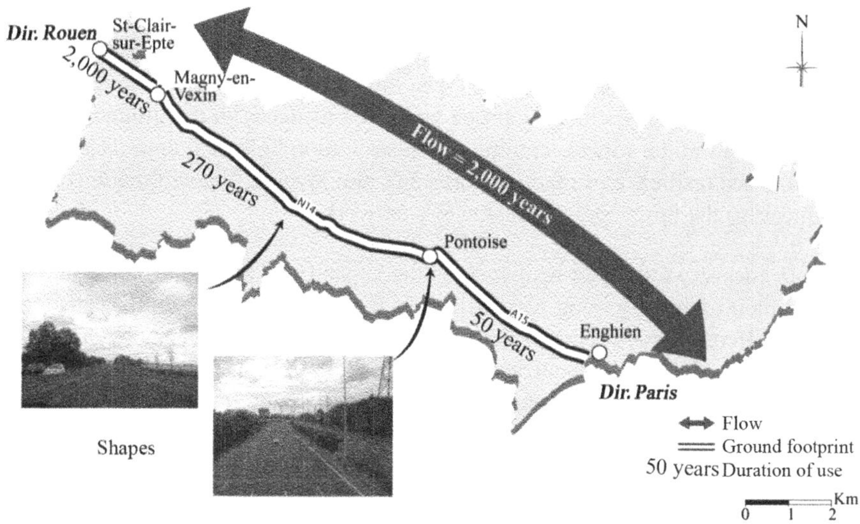


**Figure 3.18.** *Resilience of the Paris-Rouen route in the Val-d'Oise. Flow establishes coherency across road footprints of highly varied origin because of a shared contemporary road structure. Thus, every spatial dimension (flow, ground footprint, shape) corresponds to different temporal rhythms (source: Robert 2003a; photograph: S. Robert, 2019). For a color version of this figure, see www.iste.co.uk/robert/resilience.zip*

The articulation with higher spatial and temporal levels thus plays out in the shape, a level which tends to be subject to more rapid transformations. Archeological evidence shows that the transformation rhythms of shape tend to be faster than those of ground footprints. Roads are constantly re-made to ensure their serviceability, because of wear and tear and because of the development of new surfacing techniques. A single ground footprint may play host to a succession of different shapes, from an earth path to a paved road in the modern period, a tarmac surface from the 19th century on, etc. Archeological excavations, such as those carried out on the Roman *chaussée Jules César* at Courcelles-sur-Viosne (Jobic 2001) or on the Paris-Orléans road at Massy-Palaiseau (Bruant *et al.* 2014), provide evidence of significant changes to the shape of roads over a period of 50 years or less.

The maintenance of certain sites, orientations, alignments, etc., is thus possible because of temporal mismatches between the flow, ground footprint and shapes of these forms, which enable the transmission of morphological principles and consequently the reproduction of plot layouts and road networks over time. The shape of the landscape elements in these networks may vary considerably without departing from their ground footprint, as transformations on these different levels are not necessarily synchronous (Figure 3.19).

Morphogens have a long-term influence due to the fact that landscape elements may undergo transformations while retaining certain inherited morphological characteristics. Changes which might be perceived as decay in the ground footprint and shape of an initial form according to a classic morphological approach are thus essential to the resilience of road networks and plot layouts. Transformations enable adjustments to the functional dimension of a network.

In our thesis, we defined resilience as the capacity of plot and road systems to maintain their identity, exploiting conflicts of forms and drawing on the interaction between different levels (flow, ground footprint and shape), which enable the articulation of global and local levels (Robert 2003b, p. 457 and 470). Constant readjustments between the asynchronous aspects of flow, ground footprint and shape enable the articulation of new functions and inherited forms according to a non-linear temporality. Forms continue to correspond to the needs of a society by means of "an ongoing present of forms and usages"[58] (Lepetit 1995, p. 291).

58 *...un présent continué de formes et d'usages.*

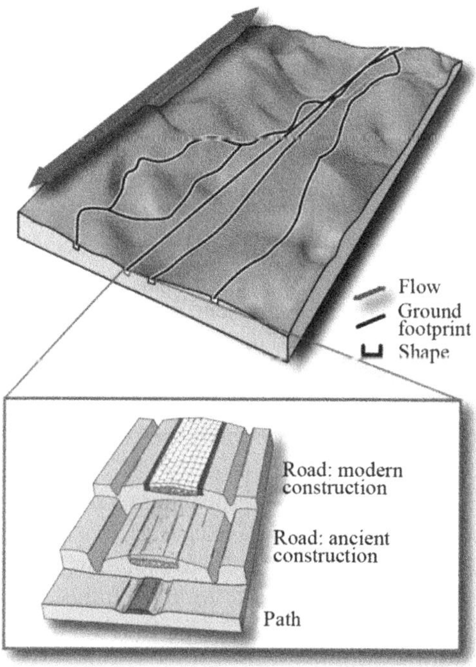


**Figure 3.19.** *The three spatial and temporal dimensions of road networks. Traffic (or flow) is carried by multiple branches (ground footprints), the aspect and serviceability (shape) of which transforms over time. Each dimension follows its own rhythms (Robert 2019)*[59]*. For a color version of this figure, see www.iste.co.uk/robert/resilience.zip*

## 3.6. Forms in a landscape: specific temporalities

As we have seen, in archeogeography, the resilience of forms is explained not only by history but also by geography, given that landscapes are seen as a hybrid of natural and social, past and present forms. In the early 2000s, G. Chouquer proposed a dissociation of the temporal aspect of forms from that of societies, describing an "internal morphological time", specific to morphological facts and transformations. The rhythms of forms may not be synchronous to the historic and social narrative, in which "dates and sequences are determined by the facts of society"[60] (Chouquer 2000, pp. 123–124).

59 Illustration inspired by Amoros and Petts (1993) who used a similar approach for fluvial hydrosystems.

60 *...ce sont les faits de société qui déterminent les dates et les séquences.*

### 3.6.1. *New temporalities in archeogeography*

G. Chouquer described five spatiotemporal modes of internal morphological time. The first two are the classic temporalities introduced by F. de Saussure: diachrony and synchrony. However, where Saussure described these temporalities as modes of observation, Chouquer presented them as temporalities, which are intrinsic to forms and, more specifically, to the relation between form and society. **Diachrony** offers the means of describing the permanence of a form and function over time, characterizing "forms [which] structure a space beyond the period of initiation, and [which] constitute a permanent framework for identical functions"[61]. **Synchrony**, on the other hand, is marked by a "rupture of form over time", and is observed "when social intervention in the environment creates a new state, coherent with the social formation in question; most planned developments correspond to this mode"[62] (Chouquer 2000, p. 126). In this case, synchrony is seen as a strict articulation of forms and social productions[63]. Any significant societal transformation should thus lead to a corresponding revision of forms; we know that this is not true in practice, due to the prevalence of mismatches between form and function in the landscape. In addition to the two "classic" temporalities, Chouquer proposed three further options, intended to reflect the non-linear temporalities observed in landscape forms.

The notion of hysteresis, borrowed from the field of ecology, forms the basis for **hysterechrony**, *hystéréchronie*:

> Hysterechrony, or a temporal mismatch between a form and social formations with the functions which they imply, describes the spatio-temporal modality of maladjustment in the morphological system. It results in morphological hysteresis, an observable lag between a 'cause' and its effect[64]. (Chouquer 2000, p. 125)

---

61 *...formes [qui] structurent l'espace au-delà de la période qui les a initiées et [qui] servent de cadre permanent à des fonctions identiques.*

62 *...rupture de la forme dans le temps [...] lorsque l'intervention sociale sur le milieu crée un état nouveau, cohérent avec la formation sociale en question ; la plupart des planifications entrent dans cette modalité.*

63 This sense is different to that used by F. de Saussure, seen more in terms of a cross-section in which elements are observed in relation to one another within the framework of a value system at a time t (section 2.3.1).

64 *L'hystéréchronie, ou décalage dans le temps de la forme par rapport aux formations sociales et aux fonctions qu'elles induisent, décrit la modalité spatio-temporelle des inadaptations du système morphologique. Elle produit l'hystérésis morphologique ou temps de latence constaté entre une "cause" et son effet.*

This notion is notably used to characterize the temporality of morphogens, which exhibit a certain temporal "mismatch", since they "exercise a persistent influence on forms, well beyond the period in which they were created and operated"[65] (Chouquer 1997, p. 21) (section 3.5.2).

Another temporality proposed by G. Chouquer is **uchrony**, *uchronie*. The term is taken from the title of a novel by Charles Renouvier, published in 1876: *Uchronie: l'utopie dans l'histoire, esquisse historique apocryphe du développement de la civilisation européenne tel qu'il n'a pas été, tel qu'il aurait pu être*[66]. In this, the author rewrites the history of European civilization according to different trajectories. Renouvier imagined what might have happened if the outcome of certain event has been different, considering the possible consequences of decisions made 1500 years earlier for 19th-century society (Renouvier 1876). In literature, the term "uchronia" is now synonymous with alternative history. In archeogeography, however, the focus is on the notion of a trajectory. The history of a system itself determines a certain number of possible scenarios. The forms created at a time t partly determine the future evolution of a landscape, meaning that its future is contained within its history:

> Uchrony, or the potentiality of a form in time, describes the spatio-temporal modality which is observed when a structure or a formal element of a structure present in a landscape at a given moment in history creates a potential which a later, new social fact draws upon at a later, unknown date in the history of the site. The modality is then expressed either through isotopy (revival of the form on the same site) or through isoclination (retrieval of the orientation of the form)[67]. (Chouquer 2000, p. 126)

Thus, forms which are present in a space at a time t, in a given historical context, have the potential to be reused and re-contribute at "a later, unknown date in the

65 *...exercent une influence persistante sur les formes, bien au-delà de leur époque de création et de leur fonctionnement.*

66 Uchronia: Utopia in History, an Apocryphal Sketch of the Development of European Civilization Not as It Was But as It Might Have Been.

67 *Uchronie ou potentialité de la forme dans le temps décrit la modalité spatio-temporelle qui se constate lorsqu'une structure ou un élément formel d'une structure imprimée dans le sol à un moment historique donné crée un potentiel qu'un fait social nouveau et ultérieur fait rejouer à un moment imprévu de l'histoire du site. La modalité s'exprime alors soit par isotopie (reprise de la forme au même emplacement), soit par isoclinie (reprise de l'orientation de la forme).*

history of the site", even if they cease to be active or disappear from the landscape in the interim (Chouquer 2000, p. 126). The successive reuses of the boundary limit at Pierrelatte, the new bridge at Cergy and the bypass at Marines, described earlier (sections 2.5.5 and 3.2.1) are all cases in which new forms are unwittingly created on an older, long-lost and long-forgotten site. Thus, the inherited context partially orients the choice of one of a multitude of possible trajectories (Figure 3.20).

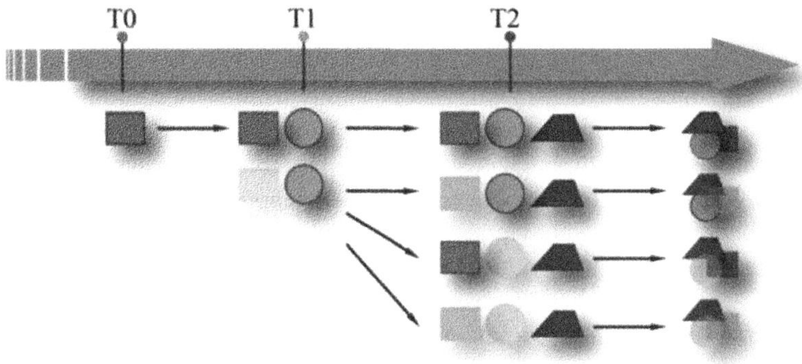


**Figure 3.20.** *Non-linear temporalities and uchrony in landscape forms. Any form inserted into a landscape at a given moment creates a potential for the future. It may be maintained or disappear, only to reappear following a period of hiatus. The past landscape produces a range of different possible scenarios for present and future landscapes (Robert 2019). For a color version of this figure, see www.iste.co.uk/robert/resilience.zip*

Finally, **taphochrony**, *taphochronie*, derived from "taphonomy" (literally "burial studies": the conditions of covering and burial of ruins and remains) is used to express the definitive disappearance of a plot layout or road network; it is "the modality of rupture in time and in the space of accumulation, by the definitive burial of a spatial state, not followed by a revival. This modality expresses irreversibility and, in this sense, is a source of historical knowledge concerning radical bifurcations in systems"[68] (Chouquer 2003b, p. 302). The disappearance of the morphological characteristics of a plot layout or road network is seen to mark the passage of the principles involved in their reproduction, and hence the end of the network as a system[69].

68 *...modalité de rupture du temps et de l'espace d'accumulation par enfouissement définitif d'un état spatial qui ne provoque pas de reprise. Cette modalité exprime une irréversibilité et est, en ce sens, source de connaissance historique sur des bifurcations radicales des systèmes.*

69 For a discussion of this term, see section 8.7.

### 3.6.2. *Time as a spiral*

While the classic "time's arrow" metaphor clearly transmits the idea of irreversibility, it is limited in its capacity to reflect the complex temporalities of forms in a landscape and of the constant interactions between past and present forms, which partly influence the future. We propose a new approach, replacing the classic arrow or cycle representations by a spiral, as used by Maurice Merleau-Ponty in *Le Visible et l'Invisible* (Merleau-Ponty 1979) and by Bruno Latour in *Nous n'avons jamais été modernes*, published in English as *We Have Never Been Modern* (Figure 3.21). In a complex approach, dynamics are considered as internal to the objects themselves. By ceasing to explain the world using a concept of nature, outside of time, and culture, in time, we cease to perceive time as a general framework, but rather as "the provisional result of the connection of beings". For this reason, Latour used a spiral, rather than an arrow or cycle, as a metaphor for time:

> Let us suppose, for example, that we are going to regroup the contemporary elements along a spiral rather than a line. We do have a future and a past, but the future takes the form of a circle expanding in all directions, and the past is not surpassed but revisited, repeated, surrounded, protected, recombined, reinterpreted and reshuffled. Elements that appear remote if we follow the spiral may turn out to be quite nearby if we compare loops. Conversely, elements that are quite contemporary, if we judge by the line, become quite remote if we traverse a spoke. Such a temporality does not oblige us to use the labels 'archaic' or 'advanced", since every cohort of contemporary elements may bring together elements from all times. In such a framework, our actions are recognized at last as polytemporal. I may use an electric drill, but I also use a hammer. The former is thirty-five years old, the latter hundreds of thousands. (Latour 1993, p. 75)

In Latour's interpretation, time is no longer seen as a linear flow, but rather as a stirring, intermingling form. Elements of irreversibility (history and the trajectory of beings) are intermingled with elements of reversibility, since history is written from a continual reshuffling of existing elements. Time is not stratified, and the way in which the complexity of phenomena is reflected in this approach establishes a permeability between eras. The past is contained in the present. In this complex approach, dynamics are not determined *a priori*; time is not seen as a linear progression or as a recurring cycle. At any point, the trajectory of a whole system may be changed by a simple bifurcation.

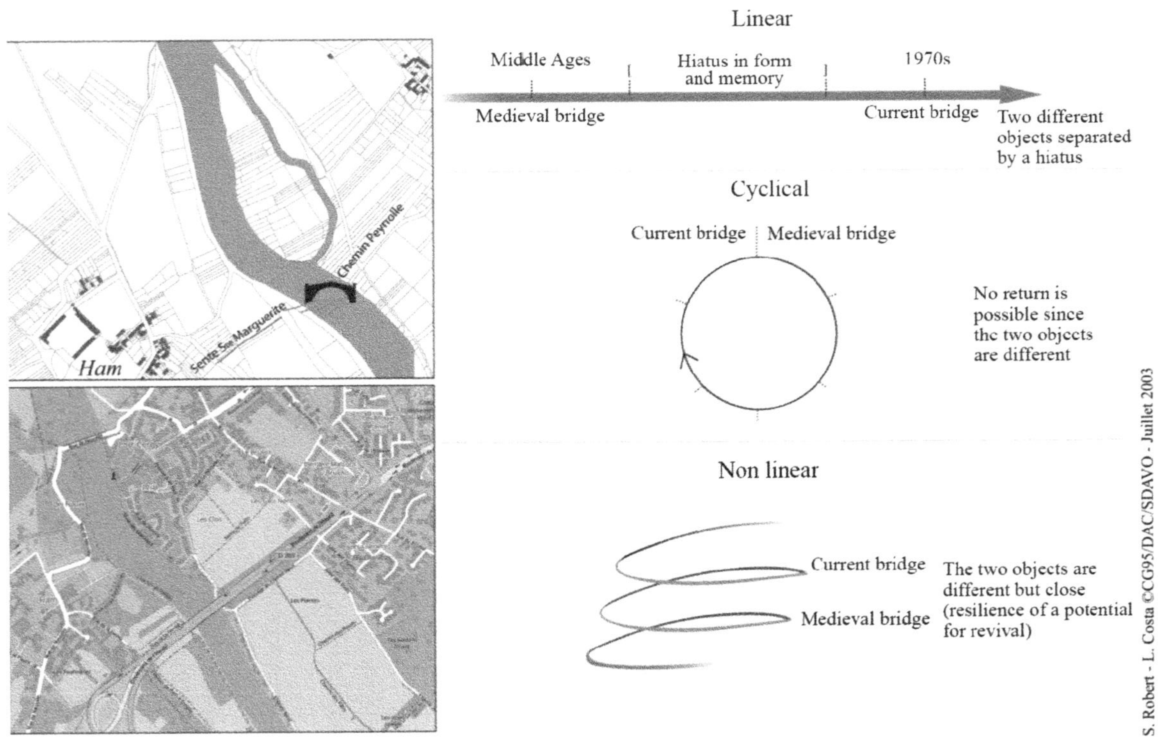


**Figure 3.21.** *Different ways of representing time. The spiral form can be used to illustrate the combination of temporalities in the present and non-linear temporalities. The current bridge at Cergy, for example, is constructed on the site of a medieval bridge which had been lost for centuries (Robert 2019, present-day map: Open Street Map). For a color version of this figure, see www.iste.co.uk/robert/resilience.zip*

## 3.7. Conclusion

From the 1990s on, increasing cross-fertilization between morphological studies, land studies and environmental studies highlighted the limitations of the semiological and stratigraphic interpretations of landscape forms developed in the late 19th century. A shift in morphological analysis methods led to the discovery of new objects of research (river corridors, formation networks, road groupings, etc.) and to a re-examination of existing objects (ancient and medieval planned elements, bocage, etc.). Plot layout and road networks came to be characterized as hybrid elements, comprising both natural and social, past and present forms. The historical explanation, which highlighted the role of planning in landscape organization, was no longer considered sufficient for characterization purposes. While landscape elements with recurring morphological characteristics (orientation, alignment, etc.) can be used to identify coherent plot or road networks, significant mismatches emerged in their dating. Empirical studies have shown that networks have the capacity to diffuse across space and time from morphogenetic structuring elements (morphogens), leading to the reproduction of morphological principles that affect new landscape elements.

Archeogeographers thus turned to complex approaches, notably the theory of self-organization, to explain the formation and reproduction of these networks; the networks themselves came to be seen as systems at the meeting-point of societies and inherited spatial contexts, made up of hybrid socionatural forms with multiple temporalities. Archeogeography has also highlighted the crucial role of circulation in the formation and resilience of plot and road networks. The reproduction of these systems is possible due to the asynchrony that exists between different spatial and temporal dimensions (flow, ground footprints and shapes), meaning that societies may reuse landscape elements over time, even if the elements in question were designed for a different purpose. Morphological analysis is thus experiencing a gradual shift away from historical-based explanations to one based on local specificities (geography), leading archeogeographers to distinguish between the temporalities of landscape forms and those of the history of societies, and to propose new, specific temporalities for the forms themselves.

Research on the subject of resilience, developed in the wake of the adoption of the self-organization paradigm, began to emerge in archeology in the 1990s as a way of studying temporal dynamics and multiscale interactions in the organization of complex systems. The revised morphological method enables closer connections between the conceptual frameworks of resilience and archeogeography, meaning that landscapes can be understood as resilient, complex, adaptative systems. Resilience has attracted significant attention in recent years, and is also a subject of debate. A detailed description of the conceptual framework for resilience proposed here is therefore essential, along with a discussion of the issues associated with its use in our present day and age.

# Part 2

# Resilience: A Tool for Understanding the Dialectics of Persistence and Change

# Introduction to Part 2

The term **resilience**, derived from the Latin *resilire*, originally means "to leap or spring back; recoil; rebound; shrink (back again)" (Oxford 2005, p. 163). This passed into Middle French as the verb *résilier*, to cancel or revoke (1082), then into English as *resile,* a term encountered in the papers of Henry VIII (1529) (Alexander 2013 p. 1261); in French, at least, it also took on a sense of "resistance" in the separate form *résiluer* (1411)[1]. The image of recoil came to dominate English usage: one of the first recorded uses of the term in a scientific context occurred in 1626, when Francis Bacon used it to denote the way in which an echo bounces back off walls (Oxford University Press 1991, p. 714). In French, the word followed a somewhat different path: on its first appearance in the Larousse dictionary in 1922, the primary definition was *le travail nécessaire pour produire la rupture d'un barreau* (the effort required to break a bar) (Larousse 1922, p. 780), referring to an experiment devised by the engineer Georges Charpy. Charpy measured the resilience of metal bars by breaking them using a pendulum impact machine. He defined the resilience of a metal as "the living resistance to rupture of an infinitely thin slice, or the work required to produce this rupture expressed in kilogrammeters per square centimeter of cross-section"[2] (Charpy 1901, p. 873). The first uses of resilience in the social sciences date from the Second World War in the context of research on attachment in children and on their capacity to survive traumatic shocks (Cyrulnik and Jorland 2012)[3].

1 See Godefroy (1881, p. 98 and 556), according to Acts 1082 and 1411.

2 *...la résistance vive à la rupture d'une tranche infiniment mince, ou le travail nécessaire pour produire cette rupture exprimé en kilogrammètres par centimètre carré de section.*

3 In France, the notion of resilience was first brought to the public's awareness through the work of the psychologist Boris Cyrulnik in the early 2000s before becoming increasingly widespread in other fields in the second half of that decade (Cyrulnik 1999).

In the early 1970s, the ecologist Crawford S. Holling developed a conceptual framework based on resilience to describe the transformation of ecosystems, building on work carried out in the field of complex systems (Holling 1973). He later referred to this framework as ecological resilience (Holling 1996). Holling's work has been disseminated by the Resilience Alliance group which he founded in 1999 with others, notably Carl Folke, the founder and director of the Stockholm Resilience Center. This center is an important focal point for research into resilience in Europe[4]. Resilience Alliance brings together researchers from the fields of ecology and the humanities, among others, running the website www.resalliance.org and the journal *Ecology and Society: Integrative Science for Resilience and Sustainability,* founded by C.S. Holling in 1997 under the name *Conservation Ecology*[5]. Research on resilience, as developed by Resilience Alliance, has led to the development of a school of thought known as resilience thinking, which aims to provide practical guidance for academic and management institutions (Walker and Salt 2006, p. XI). Resilience gained increasing importance in the academic sphere and among policymakers around the turn of the new millennium (Brown 2014, p. 3). Moving beyond the traditional fields of psychology, ecology, ecosystem sciences and environment management, resilience began to make its mark on other domains. Writing in 2014, the geographer Katrina Brown noted the role of resilience in international relations (notably in terms of managing military and terrorist threats), social ecological systems, risk and disaster management, climate change, human development (especially crisis response capacities), organizational sciences and social innovation (learning), and urban and regional development, in the case of urban resilience (Brown 2014, p. 2). In 2016, writing in the *Oxford Research Encyclopedia of Environmental Science*, Carl Folke added further areas of influence: the economy, institutions and governance, the role of the Welfare State, technology and network studies, educational research, communications, criminality and health (Folke 2016). In the field of the humanities, the journal *Resilience: International Policies, Practices and Discourses*, directed by international relations specialist David Chandler, was founded in 2013 and continues to be published by Routledge. In 2017, Chandler and the geographer Jon Coaffee edited the *Routledge Handbook of International Resilience*, exploring the potential contributions of resilience in terms of governance, adaptation to climate change, development, risk reduction, development policy, etc. (Chandler and Coaffee 2017).

Several authors have cast light on notions of resilience within their own domains, while cross-disciplinary bibliographical summaries are less common. In 2011, Patrick Martin-Breen and John Marty Anderies, on behalf of the Rockefeller Foundation, created a summary of uses of the concept of resilience over the past 50 years, touching the main fields of applications and relations with other concepts such

4 See www.stockholmresilience.org.

5 See www.ecologyandsociety.org.

as robustness, vulnerability, adaptation, etc. (Martin-Breen and Anderies 2011)[6]. In 2016, Carl Folke published a relatively thorough overview of uses of the conceptual framework of ecological resilience in disciplines connected to environmental studies (Folke 2016). Within these disciplinary or transdisciplinary overviews, authors may specify their preferred theoretical framework for resilience, although etymological and historical discussions can sometimes take the place of a true analysis of the concept itself and its theoretical implications.

Resilience is characterized in different ways by different authors and in different contexts, presented as a concept, notion, school of thought, theory[7], or even paradigm (Garcia 2015). In this book, the terms "theory" and "paradigm" will be reserved for conceptual frameworks developed in mathematics, physics and chemistry in relation to complex systems, self-organization, chaos and catastrophe theories, which form the basis for ecological resilience and which may provide a common foundation from which the different disciplines discussed here (ecology, geography and archeology) can work together. Our approach is that of the ecologist Ann Kinzig, who considers resilience as a "conceptual framework" (Kinzig 2012, p. 317). This, then, is the term used in this book to refer to ecological resilience, in the form developed by the Resilience Alliance; "concept" will be used for applications of resilience in a scientific context which do not relate to ecological resilience or which use it only partially. The term "notion" will be used in the commonly understood sense, without relating to any defined theoretical or conceptual support.

Particular attention will be paid to the framework of ecological resilience, which seems, in our view, to be the most advanced in terms of theorization and in possible applications in the field of the humanities. It notably played an important role in the first attempts to use resilience in archeology, in the context of spatial systems analysis, in the 1980s–1990s. It also builds on the same theoretical foundations as other academic currents encountered in geography, where a complex systems approach has been applied to spatial systems[8]. In our view, the conceptual framework of ecological resilience has much to offer in archeogeography, as its approach to relations between human societies and ecosystems allows us to move beyond the nature/culture dichotomy established by modern philosophy. The use of resilience in the ecological sciences forms part of a broader shift, away from the study of "pristine" ecosystems, untouched by man, and toward the study of social-ecological systems (SES). These are defined as complex, integrated systems,

---

6 Part of the *Building Climate Change Resilience Initiative*, financed by the Rockefeller Foundation from 2007.

7 For example, see the chapter *In Quest of a Theory of Adaptive Change*, in which the authors express a desire to create a new theoretical approach to understanding change (Gunderson and Holling 2002, p. 10).

8 For the French context, see, for example, Pumain *et al.* (1989) or Sanders (1992).

in which human and nature form part of the same dynamic system. SES are characterized by dependency and feedback relationships (Berkes and Folke 1998; Folke *et al.* 2010, tab. 1; Martin-Breen and Anderies 2011, p. 15). Taking this approach, it is theoretically impossible to study human societies and ecosystems in isolation. The concept of co-evolution, introduced in the 1990s, marked a first shift in the understanding of relations between man and environment; social ecological systems go much further in terms of integration (Norgaard 2009). The following chapters provide a detailed description of the conceptual framework of ecological resilience, and an examination of its applications in relation to spatial systems and landscapes.

# 4

# Ecological Resilience as a Systemic Property of Social-ecological Systems

In 1973, in an article now considered as one of the founding documents of thought on the subject, Holling (1973) referred to resilience in modeling ecosystems. In a 2006 text, looking back over his research career, Holling (2006) noted that this choice marked a shift from research into simple systems, associated with a high degree of certainty, to research into complex systems where uncertainty plays a significant role (p. 2)[1].

## 4.1. The roots of resilience

In the 1970s, C.S. Holling examined the way in which organisms are affected by their environment, but also the way in which they themselves affect the environment via feedback effects. He showed that these interactions may result in the creation of new structures, differing according to scale. At the time Holling was writing, ecosystems were widely considered to possess stable equilibrium states[2]. Examining studies of different ecosystems (insects, birds, mammals, fish, etc.), Holling hypothesized that each of these systems possessed several stable states, and that this variability is necessary to their survival. In 2006, looking back over his work in the 1960s and 1970s, the author noted his desire to develop "a different eye to see and a different theory" to identify the consequences of this discovery (Holling 2006, p. 6).

1 In this text, C. Holling reviewed the five works which, in his view, best illustrated the development of his thought (Holling 1965, 1973, 1986, 1992; Gunderson and Holling 2002).

2 A system is said to be in equilibrium when the forces acting upon it balance out. Following a disturbance, the system is subject to a force which pushes it back to its state of equilibrium (Taillet *et al.* 2013, p. 251).

C. S. Holling provided illustrations of resilience, notably in the case of predator–prey systems in which two populations interact. In the absence of predators, the population of a prey species will tend to increase in a constant manner, assuming there is no competition. In the absence of prey, the predator population will tend to decline due to the lack of food. The two populations self-regulate through trial and error. In an article published in 1973, Holling represented interactions within this type of system in a phase space, used in physics to represent the evolution of a system as "a trajectory showing the evolution of one variable as a function of another"[3] (Sanders 1992, p. 186) (Figure 4.1).

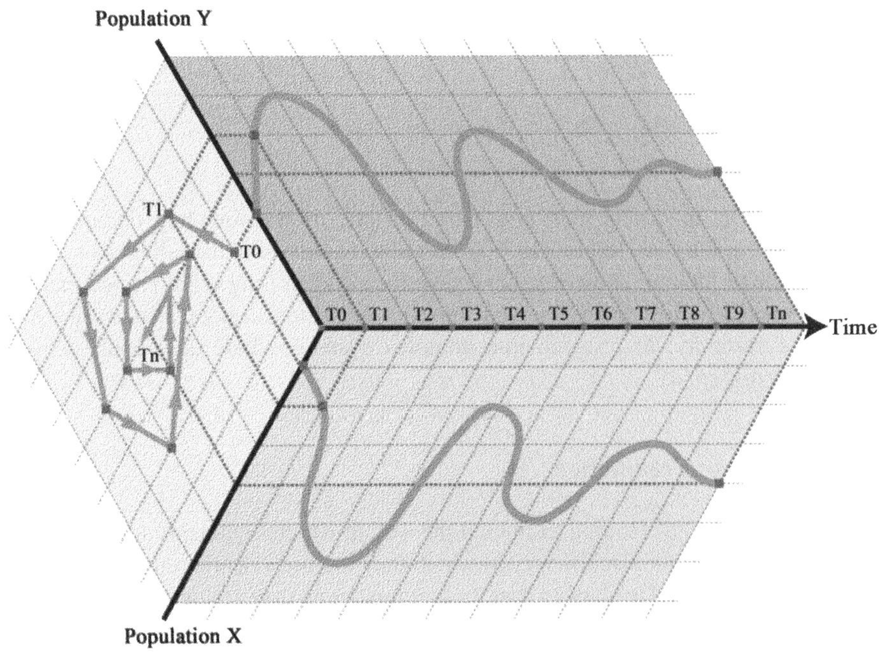


**Figure 4.1.** *Representation of the numerical evolutions of two interdependent populations (e.g. predators and prey) in a phase space. The rectangular faces show the evolution of each population separately. The square face shows the trajectory of the system formed by the interaction of the two populations. The green dotted lines indicate the correspondence between the two curves for T0, T1 and Tn (adapted from Holling (1973, fig. 1) by S. Robert, 2019). For a color version of this figure, see www.iste.co.uk/robert/resilience.zip*

The **variable space** is the space that describes all of the points defined by the values taken by variables when represented, for example, in a graphical form. The

3 *...une trajectoire montrant l'évolution d'une variable en fonction d'une autre.*

variable space toward which the trajectories of dynamic systems tend is known, in physics, as an attractor. The **attractor** of a system is thus "the state towards which the system tends, in the absence of disturbances, whatever its initial situation"[4] (Sanders 2020, p. 141). An attractor may take different forms: a fixed point, a one-dimensional curve, a surface of higher dimensions, or even a fractal object, known as a "strange attractor", in the case of chaotic systems with unpredictable trajectories (Sanders 2017, p. 35)[5]. In the example shown in Figure 4.1, the attractor is a fixed point. The space within which all trajectories converge toward the same attractor is known as a **basin of attraction**. To use a metaphor particularly suited to the contexts of geography and archeology, the attractor may be defined as the state "towards which all trajectories situated in the same basin of attraction converge, just as all surface water in a valley will descend to the river at the bottom of that valley"[6] (Encyclopedia Universalis 2001, p. 145).

To illustrate the development trajectory taken by organisms among several possible trajectories, C.S. Holling used the metaphor of a ball moving through a landscape of peaks and valleys (Holling 1986, p. 72, 1996, p. 35)[7]. The bottom of the valley is the attractor toward which the system tends, while the sides of the valley represent the basin of attraction (Figure 4.2). In this type of representation, the faster the ball returns to the bottom of the valley, the greater the stability of the equilibrium state. The summits of the peaks represent situations of unstable equilibrium, where the system may tip into another valley (another attractor) with a complete change of regime, that is, of function and structure. The system moves from one basin of attraction to another (another valley), partly under the effect of internal processes and partly as a result of external events. Passages from one valley to another are known as "transitions" or "bifurcations" (section 4.5).

The notion of the basin of attraction is at the heart of the definition of resilience proposed by C.S. Holling in 1973. Up until that point, system dynamics were essentially studied in relation to the notion of stability, "the property of a body,

4 ... *l'état vers lequel se dirige le système, en l'absence de perturbation, quelle que soit sa situation initiale.*

5 See Holling (1973) and Figure 4.2.

6 ...*vers lequel convergent toutes les trajectoires situées dans le bassin d'attraction correspondant, de même que tous les ruissellements d'une vallée aboutissent dans la rivière qui coule dans le fond.*

7 This image is also used in the "epigenetic landscape" created by the biologist Conrad Hal Waddington in 1957, to illustrate cellular differentiation based on the notion of sensitivity to initial conditions (Bourgine and Lesne 2006, p. 305). Cristina Aschan-Leygonie used this idea in an article published in 2000 (Aschan-Leygonie 2000, Figure 1b).

disturbed from its state of equilibrium, to return to that state"[8] (Larousse 1922, p. 992). Holling chose, instead, to focus on the persistence of relations in a system:

> There is another property, termed resilience, that is a measure of the persistence of systems and of their ability to absorb change and disturbance and still maintain the same relationships between populations or state variables. (Holling 1973, p. 14)

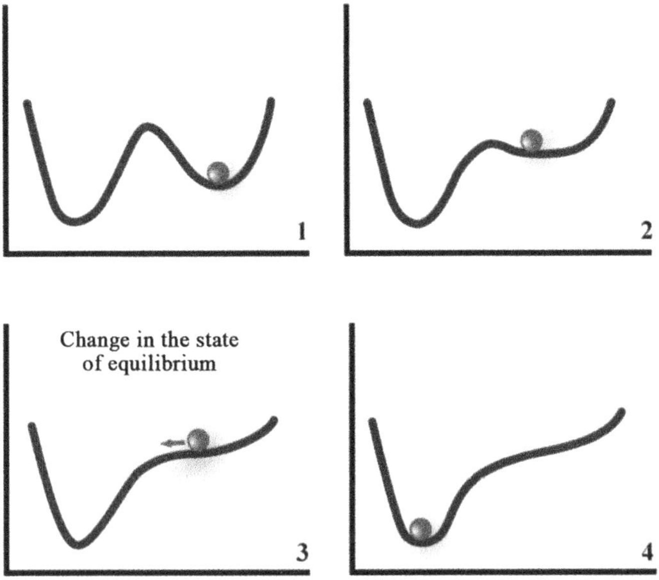


**Figure 4.2.** *Illustration of the notions of attractors and basins of attraction. The red ball indicates the situation of a system. In step 3, the system is moving from one basin of attraction to another (another valley) (adapted from Holling (1996) by S. Robert, 2019). For a color version of this figure, see www.iste.co.uk/robert/resilience. zip*

Scheffer *et al.* later built on this idea, defining resilience in the following terms:

> Following Holling, we here use the term "resilience" to refer the size of the valley, or basin of attraction, around a state, which corresponds to the maximum perturbation that can be taken without causing a shift to an alternative stable state. (Scheffer *et al.* 2001, p. 591)

Thus, the different parameters used to measure resilience relate to the characteristics of the basin of attraction. Holling defined these parameters in 1986:

---

8 *Propriété que possède un corps, dérangé de son état d'équilibre, de revenir à cet état.*

> The size of the stability domain of residence, the strength of the repulsive forces at the boundary, and the resistance of the domain to contraction are all distinct measures of resilience. (Holling 1986, p. 76)

Walker *et al.* defined three notable attributes. First, the **latitude** or width of the basin of attraction (noted L in Figure 4.3) corresponds to "the maximum amount the system can be changed before losing its ability to recover". For Walter *et al.*, the wider the basin, the more states a system may pass through without reaching a definitive tipping point. The second attribute, **resistance** (noted R), indicates the ease or difficulty with which the system may change. The deeper the basin of attraction (defined by the ratio R/L), the greater the perturbation or force required to move the system away from the attractor. Third, **precariousness** (Pr) defines the trajectory of a system and its proximity to a tipping point after which recovery is difficult or impossible (Walker *et al.* 2004, pp. 6–7).

C.S. Holling considered that resilience was more suitable than stability as a starting point for the study of complex adaptive systems, in which highly variable behaviors, far from the state of equilibrium, are a significant factor in ensuring the survival of the system (Holling 2006, p. 2).

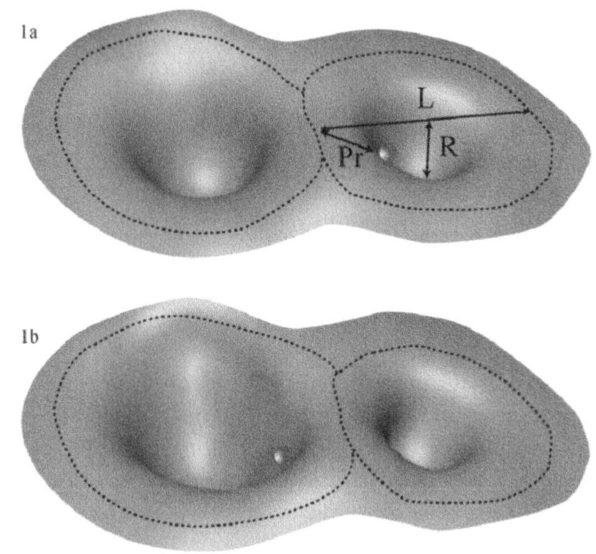


**Figure 4.3.** *Representation of a stability landscape featuring two basins of attraction, showing three aspects of system resilience: latitude (L), resistance (R) and precariousness (Pr). Image 1b shows changes to the stability landscape following the contraction of the original basin of attraction and the expansion of the alternative basin (adapted from Walker* et al. *(2004, p. 4) by S. Robert, 2019). For a color version of this figure, see www.iste.co.uk/robert/resilience.zip*

## 4.2. Resilience versus stability: the dynamic role of perturbations

From 1973, C.S. Holling indicated a clear opposition between resilience and the concept of stability. However, over the following years, resilience tended to be assimilated to stability, including within the field of ecology, notably after the publication of an article by the ecologist Stuart L. Pimm in the journal *Nature* in 1984. In this, resilience was defined as the calculation of the speed at which the characteristics of a complex, stable ecosystem return to a state of equilibrium following a perturbation, and was calculated in units of time (Pimm 1984, p. 322). In the 1990, Holling considered that this definition of resilience as a calculation of resistance to perturbations and of the speed of return to equilibrium had become the most common; he labeled this interpretation "engineering resilience", as opposed to "ecological resilience" (Holling 1996). Engineering resilience became particularly dominant in management solutions for the field of risk management and development, as it can be modeled in a relatively simple mathematical form in which perturbations and the modes of return to the stable state are considered to be predictable. Thus, societal responses to perturbations should focus on increasing resistance, elasticity and the extent of the basin of attraction of the system (Martin-Breen and Anderies 2011, p. 44). In Holling's view, while this engineering resilience produces short-term results, it can also lead to a reduction in long-term resilience (Holling 1996, p. 37). The confusion of the two types of resilience played a major role in critical debates from the 2000s on (section 5.2).

The perspective of ecological resilience, which focuses primarily on the persistence of relations, is thus radically different from that of engineering resilience; in the ecological approach, it is the capacity of a system to absorb change while continuing to exist, rather than to return to a given state, that is in question (Holling 1973, p. 1 and 17). A system with low stability may, in fact, be highly resilient. Holling (1973) cited the example of a caterpillar known as the spruce budworm, the forest populations of which are highly unstable, but "it is because of this instability that it has an enormous resilience" (p. 14). One new aspect introduced in ecological resilience relates to the role of perturbations. Far from leading to the destruction of a system, perturbations – such as forest fires – can contribute to the maintenance of ecosystems (Holling 1978, p. 10). Holling (1978) developed the original idea that "resilience is a property that allows a system to absorb and *utilize* (or even benefit from) change" (p. 11).

There are thus two radically different ways of understanding change. In engineering resilience, which is particularly widespread in the operational domain, the effects of perturbations are minimized, with a focus on returning to stability. In ecological resilience, perturbations act as a motor of resilience by forcing systems to adapt. The former understanding of the dynamics of ecosystems, in which species evolve toward a climactic point with a stable, durable habitat obtained through a self-reproducing assembly, is called into question. In the stability-centered approach, certain events (fires, storms, pests, etc.) are considered exogeneous to the dynamics of the system (Holling 1986, pp. 78–79). From the 1980s on, building on studies carried out in collaboration with ecologists, economists, mathematicians, etc., Holling began to develop a dynamic approach to resilience in the form of the adaptive cycle.

## 4.3. A dynamic approach to organization: the adaptive cycle

C.S. Holling identified a need to move beyond a purely functional vision and to reason in terms of organization: the elements of ecosystems are interconnected into sub-systems, which themselves form part of broader groupings (Holling 1986, p. 74). Holling defined the systems he observed as "communities of organisms in which internal interactions between the organisms determine behavior more than do external biological events" (Holling 1986, p. 77). In different system types (forest fire propagation, destructive insect invasions, etc.), he identified different operating speeds: fast, slow, or intermediate, depending on the variable under observation[9]. In his case studies, the broadest spatial scale was determined by the dispersion distance of the most mobile of the observed variables (such as the dispersion space of an insect species), while the slowest time scale was determined by the longest living elements (such as trees, in comparison with leaves or needles, as presented in Holling 1986, pp. 77–78). In some cases, the landscape is thus presented as the scale at which the slowest rhythms are observed in terms of vegetation (over centuries, or even millennia) and as the largest spatial scale (from 10 km up to 1,000 km or more; Holling 1992, p. 475; Gunderson *et al.* 2002, p. 68). Holling proposed that the dynamics of the evolution of ecosystems could be seen as an adaptative cycle, taking account of different rhythms (Holling 1986, p. 95, 1992, p. 481; Holling and Gunderson 2002, pp. 32–47) (Figure 4.4).

The **adaptive cycle** takes the form of a lazy eight, representing the succession of four functions. These were presented in different publications under different names, before a fixed set of letters, r, K, Ω and α, was fixed upon in 2002. The **r exploitation or growth phase** marks the beginning of the cycle. "Opportunist" species, "r-strategists", maximize growth in unstable environments, notably zones that have recently experienced perturbation. These pioneering species have a high reproductive power, short lifespan, small size, high dispersion and high resistance to extreme conditions. They grow rapidly, in physical conditions which may be extreme. They progressively allow entry to robust, more competitive species, the "K-strategists". The arrival of the latter marks the start of the **K conservation phase**. "K-strategists" have a low reproductive power, longer lifespan, a lower rate of spatial dispersion, a large size and efficient competitive capacities. K-strategists evolve in a

9 See, for instance, Holling (1986, Table 1, p. 85).

relatively stable manner, and the K phase is characterized by the accumulation and storage of energy and matter (Holling 1986, p. 95; Holling and Gunderson 2002, pp. 33–35). In the first version of the adaptive cycle in 1986 (Figure 4.4), C.S. Holling (1986) associated the K conservation phase with the notion of climax (p. 95) (Figure 4.5). This represents an ideal state to be attained and maintained, in a vision not far distant from that of a finished, initial, ideal state which we explored in our earlier presentation of landscape research.

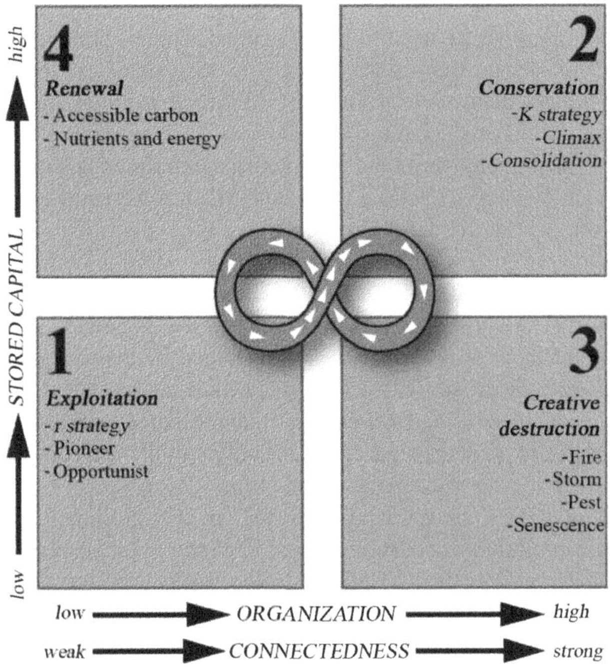


**Figure 4.4.** *One of the first representations of the adaptive cycle, with the four functions of exosystems organized according to their relations with the quantity of stored capital and the degree of connectedness in these systems. The intervals between the white arrows give indication of the speed of change (short interval: slow change, long interval: rapid change; Holling 1986, p. 95, by S. Robert, 2019). For a color version of this figure, see www.iste.co.uk/robert/resilience.zip*

However, Holling chose to move away from this vision by proposing another type of evolution, expressed through two new functions. He suggested that the success of K-strategists in competition with other species and in the appropriation and accumulation of energy, space and nutrient resources makes the system more vulnerable to external agents, due to an overly strong connection between some of its constituent elements. In an over-connected network, change can occur more quickly. An increasing dependency between man and certain natural resources, for instance, would make both more vulnerable. The homogenization of tree species in a managed forest makes the whole forest more vulnerable as it is more sensitive to destructive insect attacks (Holling 1986, p. 102). Increased interdependence between the elements of a system results in spatial homogenization and could cause the whole system to fall into a "rigidity trap" (Gunderson *et al.* 2002, p. 95). The system may then be subject to brutal transformation. Accumulated substances would be freed to fulfill a third function, that of **creative destruction**, an expression borrowed from the economist Joseph Schumpeter. This function is also referred to as the **release** or **Ω phase** (Gunderson and Holling 2002, p. 34). During the release phase, tightly interconnected resources, which have become increasingly fragile, are suddenly released[10]. By means of decomposition processes, the stored capital reintroduces energy into the system (as in the case of soil regeneration following a fire). Holling thus associates the notion of destruction with that of creation in an innovative manner.

This phase gives rise to a fourth function, the **renewal, reorganization** (Holling 1992, p. 481) or **α phase** (Holling and Gunderson 2002, p. 35). Resources are reorganized into a new system to gain new opportunities. This phase is considered to be essential: "the engine of variety and the generator of new experiments within each level" (Gunderson *et al.* 2002, p. 74). For example, this is the moment when technical innovations may occur. In 1992, Holling assimilated resilience to these two new phases:

> Those cycles are organized by four functions: exploitation, conservation, release, and reorganization. Stability and productivity are determined by the exploitation and conservation sequence. Resilience and recovery are determined by the release and reorganization sequence. (Holling 1992, p. 450)

Resilience thus corresponds to the capacity of a system to reorganize in case of significant perturbations (see section 6.2 for examples of applications of the adaptive cycle).

In the most common representation of the adaptive cycle (Figure 4.5), the y axis represents the evolution of the potential to accumulate resources (such as biomass and nutrients), while the x axis represents the evolution of the degree of connectedness between variables. There is a continuous progression from r to α. It is slow from the exploitation phase, r, toward the conservation phase K, but very fast when the system moves from K toward the phase of liberation or creative destruction Ω, and between Ω and the reorganization phase α. Progression is also

10 Resilience Alliance, accessed May 24, 2019.

fast between α and a new exploitation phase r. In the 2002 version of the diagram, the "8" is no longer closed: there is an exit on the left, illustrating cases where a shift to a less productive or less *organized* system is possible (Holling and Gunderson 2002, p. 34).

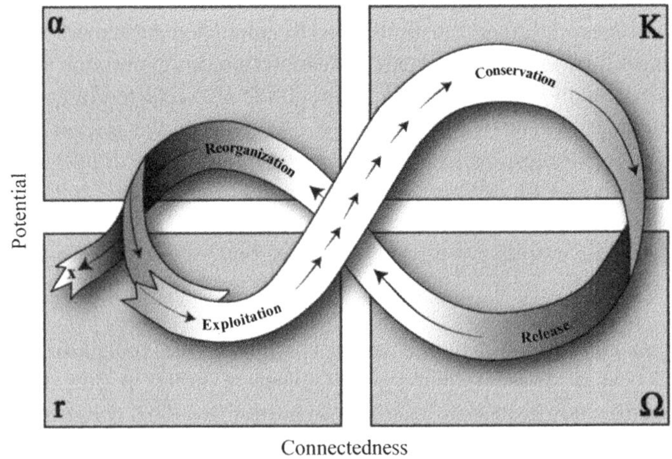


**Figure 4.5.** *The most common current representation of the adaptive cycle, in the form of a non-closed ribbon. Progression is continuous from r toward α. It is slow (small arrows) between r and K, then faster (long arrows) between and Ω, then between Ω and α (adapted from Holling and Gunderson (2002, p. 34), by S. Robert, 2019). For a color version of this figure, see www.iste.co.uk/robert/resilience.zip*

## 4.4. The panarchy model

### 4.4.1. *Development of the model*

The model of the adaptive cycle, represented by a closed "8" in its initial 1986 form, is relatively cyclical in appearance. Although the passages between different sequences are marked by different rhythms, the relation between the phases is essentially one of succession. In the late 1990s and early 2000s, the model developed further to create a finer connection and articulation of the spatial and temporal levels in systems (Holling 2006, p. 10), notably through the collective multidisciplinary work carried out at the Beijer Institute of Ecological Economics, where researchers aimed to develop a theory integrating the traditionally separate approaches of ecology, economics and the social sciences (Holling 2006, p. 11 and 13)[11].

In economics, the adoption of the paradigm complexity and the appearance of works such as *Governing the Commons* by Elinor Ostrom, published in 1990, created a field of research into which resilience could be integrated with relative ease (Ostrom 1990)[12]. Using examples observed over long periods of time, the American economist developed a theory of self-organized collective action, showing that individuals are capable of cooperating and self-organizing to maintain the natural resources upon which they depend, without recourse to a third party (State or firm). In her book, Ostrom draws on the concept of **robustness** which, like resilience, includes notions of system trajectories, learning and adaptation. P. Martin-Breen and J.M. Anderies defined robustness, like resilience, as "the capacity of a system to continue to function given external shocks" going on to specify that robustness differs from resilience in its application to a fixed system and a fixed set of external shocks. It tends to be observed over a small time scale, and mostly concerns designed systems with a given goal (Martin-Breen and Anderies 2011, p. 14 and 41).

In the early 2000s, the program developed by Holling *et al.* resulted in the publication of several key works in thought on resilience, including *Panarchy: Understanding Transformations in Human and Natural Systems*, which presented the general and cross-disciplinary findings of the Resilience Alliance (Gunderson and Holling 2002). Lance Gunderson, Garry Peterson and Crawford Holling presented a broader view of the adaptive cycle, that of **panarchy**, defined as follows on the Resilience Alliance website:

> A panarchy is a model of linked, hierarchically arranged adaptive cycles that represents the cross-scale dynamic interactions among the levels of a system and considers the interplay between change and persistence[13].

The authors indicated that the term "panarchy", referring to the Greek god of nature, Pan, was chosen over that of "hierarchy", which they felt implied an

11 The Beijer Institute of the Swedish Academy of Sciences held regular meetings, bringing together economists and naturalists to work on developing a common basis for research, from 1993 onward, notably on the island of Askö. www.beijer.kva.se, consulted April 2, 2018; see also Gunderson *et al.* (1997).

12 E. Ostrom has been active in the field of resilience studies since the mid-2000s, notably through her work with the Beijer Institute of the Swedish Academy of Sciences and the Stockholm Resilience Center (Folke 2016, p. 6).

13 Resilience Alliance – Glossary: https://www.resalliance.org/glossary, consulted September 4, 2018).

excessively top-down approach (Gunderson and Holling 2002, pp. 73–74)[14]. The panarchic model articulates three levels of temporal rhythms and spatial domains, each with their own rhythm and adaptive cycle (Figure 4.6).

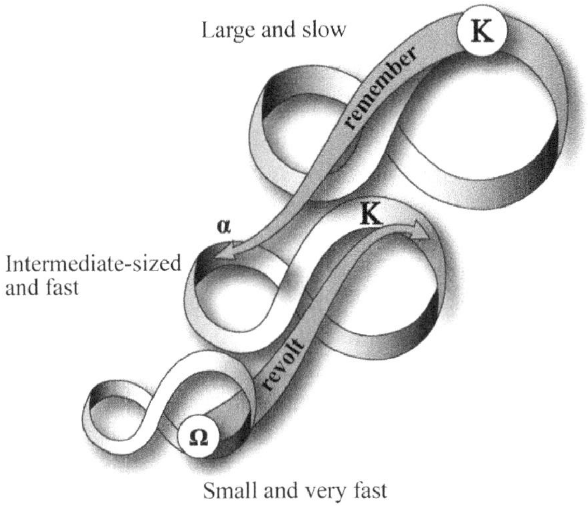


**Figure 4.6.** *Representation of panarchy, articulating three levels of adaptive cycles (adapted from Gunderson and Holling (2002, p. 75), by S. Robert, 2019). For a color version of this figure, see www.iste.co.uk/robert/resilience.zip*

In the case of ecosystems, the **large and slow level** may describe elements operating over spaces up to hundreds of kilometers and time periods of up to a millennium (e.g. landscape, climate, geomorphological and biogeographical processes, etc.). The **intermediate size and speed level** describes elements from tens of meters to several kilometers over short periods of time (such as fire, storms or pest attacks in a forest environment). The **small and fast level** corresponds to elements on a smaller scale, such as the leaves or needles on a tree (Holling and Gunderson 2002, p. 69). The common metaphorical representation of panarchy used by the Resilience Alliance illustrates the three levels using adaptive cycles, which are proportional to their extent (Gunderson and Holling 2002, p. 75) (Figure 4.6). A second representation of panarchy, without the intermediate level, also exists (Gunderson *et al.* 2010, p. 8) (Figure 4.7).

14 The authors made no reference to the definition of panarchy as a political approach. In 1860, the botanist and essayist Paul Émile de Puydt used the term to describe the right of any individual to affiliate him or herself with any (or no) type of government, whatever their origins, geographic localization, etc. This school of thought has enjoyed something of a resurgence since the 2000s; see, for example Tucker and Gian Piero de Bellis (2015).

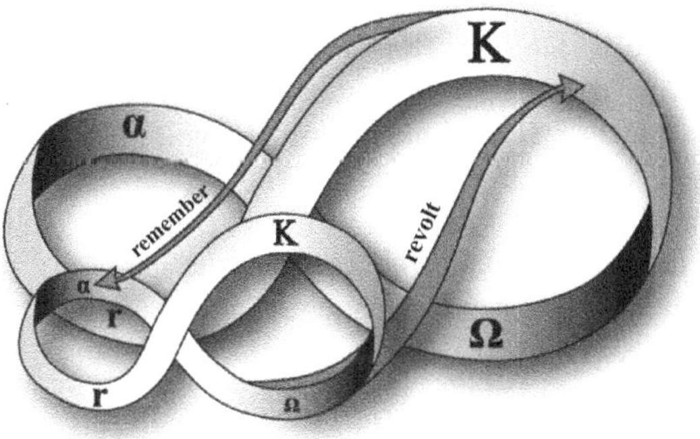


**Figure 4.7.** *Representation of panarchy articulating two levels of adaptive cycles (adapted from Gunderson* et al. *(2010, p. 8) by S. Robert, 2019). For a color version of this figure, see www.iste.co.uk/robert/resilience.zip*

In 2004, Walker *et al.* gave a more detailed explanation of panarchy (*Pa* in Figure 4.8) looking at the way in which the states of a system (notably the three attributes of latitude, resistance and precariousness) may be influenced by the states and dynamics of systems or sub-systems on higher or lower levels of the focal scale. Those situated on a finer scale may have a direct effect on the system, while those on a coarser scale may modify the stability landscape. The authors combined the representation of panarchy with the marble-in-a-valley metaphor (Walker *et al.* 2004, p. 5 and 7) (Figure 4.8).

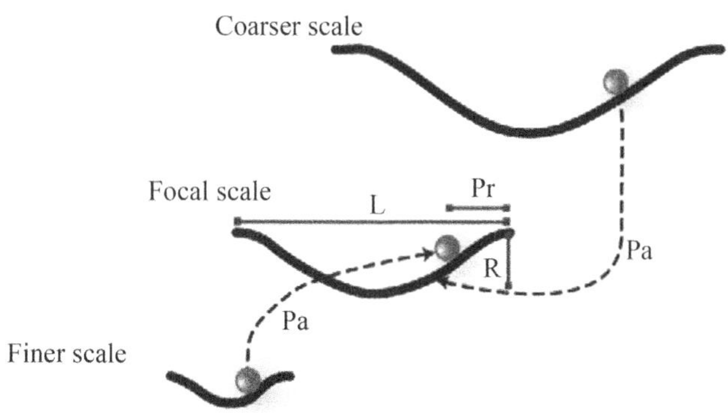


**Figure 4.8.** *Another way of representing panarchy (Pa): "The fourth aspect of resilience in relation to a stability landscape – Panarchy (Pa); the influence of the states of the system (including where they are in their adaptive cycles) at scales above and below the focal scale, which affects the other three aspects (Figure 1) by impacting the system directly (from the finer scale) or changing the stability landscape (from the coarser scale)" (adapted from Walker* et al. *(2004, p. 5) by S. Robert, 2019). For a color version of this figure, see www.iste.co.uk/robert/resilience.zip*

The originality of the panarchic model lies in the introduction of connections between different levels. Holling, Gunderson and Peterson identified two types of connections that are particularly important, notably during periods of change: "revolt" and "remember". The **revolt arrow** in Figures 4.6 and 4.7 indicates that events occurring in cycles of intermediate and fast speed may influence slower levels, pushing them into the creative destruction phase, Ω. Holling *et al.* cited the example of local campaign groups who, by opposing the use of insecticides, caused a change in regional forest management policies, which actually increased vulnerability (Gunderson *et al.* 2002, p. 76). In Holling's view, one of the innovations of the panarchic model lies in recognizing the fact that the small, fast level can transform the large, slow level. Panarchy, thus, accords a greater role to the individual scale (Holling 2006, pp. 20–21).

In contrast, the **remember arrow** may have an impact during the renewal phase α, facilitating reorganization by enabling the system to draw on the potential accumulated at other levels. In the case of forest fires, for example, the renewal phase may be helped along by the existence of a seed bank, physical structures and surviving species, which form a biotic legacy accumulated over the long-term period of growth in the forest. For the authors, it seemed as if "this connection draws upon the accumulated wisdom and experience of maturity – hence the choice of the word remember" (Gunderson *et al.* 2002, p. 76).

Panarchy involves an articulation between levels ranging from a small scale (a centimeter or a day) up to a much larger scale (hundreds of kilometers, or millennia), or from the individual up to the planet (Holling 2006, p. 22). The "revolt" and "remember" arrows create connections between these different levels, causing accelerations or decelerations in cycles on a higher or lower level. They may reinforce the resilience of a system, or accelerate its collapse. In the panarchic model, operation on different scales is seen in terms of articulations and interdependencies, rather than as a nested structure. The authors summarize the model as follows:

> The fast levels invent, experiment, and test; the slower levels stabilize and conserve accumulated memory of past successful, surviving experiments. The whole panarchy is both creative and conserving. The interactions between cycles in a panarchy combine learning with continuity. (Gunderson *et al.* 2002, p. 76)

Alongside the possibility of illustrating the articulation between different temporalities and spaces in system dynamics, it is this focus on the complex relationship between transformation and persistence within the concept of panarchy that makes it particularly attractive; it corresponds closely to our own observations in archeogeography since the 1990s. It echoes the articulation of flows, ground footprint and models which we put forward in 2003 (section 3.5.4), and bears a clear resemblance to our spiral diagram (section 3.6.2).

### 4.4.2. *Panarchy and social systems*

From the late 1970s, C.S. Holling noted that the concept of resilience could be applied to human systems as, in his view, they possess some of the same characteristics seen in ecosystems. They feature multiple basins of attraction, a fact which may result in sudden transitions, but also facilitates responses to uncertainty. Societies exhibit the primary characteristics – surprise, uncertainty and unpredictability – which has led Holling to use the concept of resilience in studying ecosystems (Holling 1978, p. 36).

In the 1980s, Holling proposed an analogy between the functions featured in his adaptive cycle and those used in economics, technology, psychology and institutional studies (Holling 1986, p. 104, tab. 2). In 1998, Carl Folke and Fikret Berkes published a book in which they explored the links between resilience in ecosystems and characteristics developed by local societies, including technologies, knowledge of their environment, property laws, conflict with higher level institutions, etc. (Folke and Berkes 1998). The book *Panarchy: Understanding Transformations in Human and Natural Systems*, mentioned earlier, marked an important step in attempts to adapt the ecological idea of resilience to create an integrated theory of ecological and social systems. The authors aimed "to identify how economic growth and human development depend upon joint attributes of ecosystems and institutions" (Gunderson and Holling 2002, p. XXI); most of the case studies presented involved the application of key elements of resilience to social systems.

In *Surprises and Sustainability: Cycles of Renewal in the Everglades*, for instance, Gunderson *et al.* (2002) developed a case study illustrating the possibility and interest of adapting the four phases of the adaptive cycle and the panarchy model to the interpretation of the dynamics of social systems, in tandem with ecological systems. They proposed an interpretation of the four main phases of water management policies in the Florida Everglades over the course of the 20th century. Each phase was considered as a different adaptive cycle, emerging after a renewal phase, itself following a crisis caused by flooding or drought. Crisis situations led to the creation of new management structures and types, corresponding to the start of a new regime of stability (Figure 4.9). This example illustrates an aspect which, according to Resilience Alliance researchers, sets human systems apart from ecological systems. Human systems have the capacity to innovate and to generate new types of social structures:

A unique property of human systems in response to uncertainty is the generation of new types of social structures. Novelty is key in responding to surprises or crises. Humans are unique in that they create novelty that transforms the future over multiple decades to centuries. (Gunderson *et al.* 2002, p. 325)

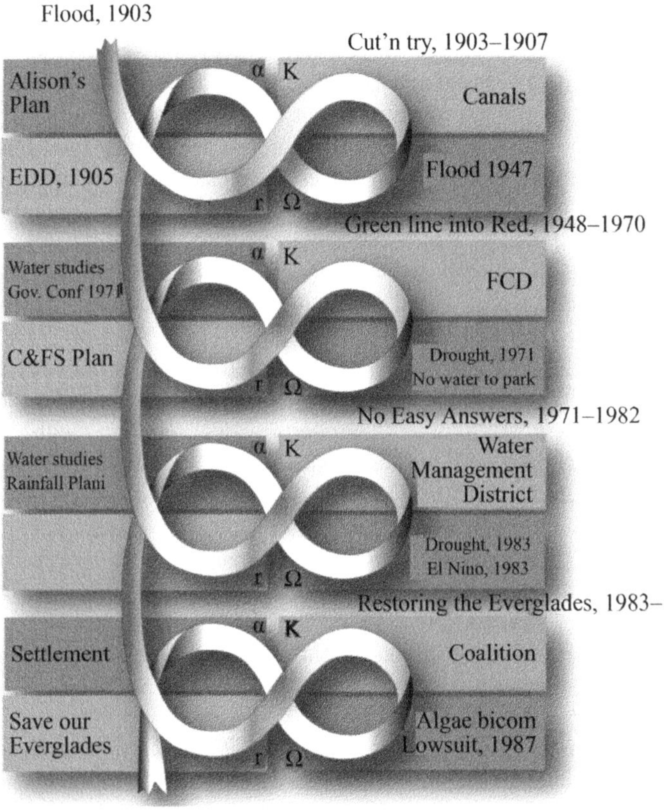


**Figure 4.9.** *Interpretation of the four main phases of development in the Florida Everglades between 1903 and the present day, seen as an iteration of four successive adaptive cycles. Significant floods and droughts forced public authorities to alter their development policies, and may be seen as phases of creative destruction (Ω phase) (adapted from Gunderson* et al. *(2002, p. 316) by S. Robert, 2019)*[15]*. For a color version of this figure, see www.iste.co.uk/robert/resilience.zip*

15 The original illustration includes a flood in 1903, although in the text the authors talk of a drought at this time. The error has been corrected here with L. Gunderson's permission.

In the panarchic model which they developed on the basis of the Everglades study (Figure 4.10), the authors proposed an articulation of three interacting levels, each with their own adaptative cycle. Since the 1920s, the large, slow level, which takes the form of federal resources (finances, expertise, etc.) has been mobilized by local authorities (intermediate level) to limit environmental crises in the everglade ecosystem (small, fast level). In this case, the proposed model was extended beyond the general panarchy model (Figures 4.6 and 4.7): revolt arrows start in both the fast and intermediate levels, illustrating the way in which local policies can act on federal resources, while remember arrows start in both the slow and intermediate levels. This example highlights the importance of the intermediate level: in this case, local policies.

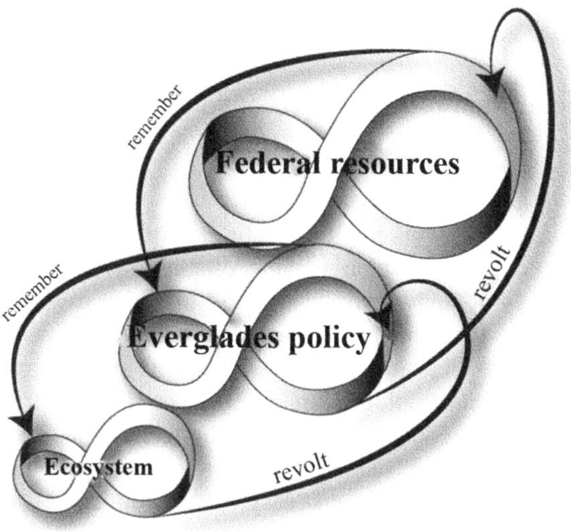


**Figure 4.10.** *Proposed representation of panarchy in the Everglades case study by L. Gunderson, C.S. Holling and G. Peterson (adapted from Gunderson* et al. *(2002, p. 331) by S. Robert, 2019)*

COMMENT ON FIGURE 4.10.– *A remember arrow, directed toward the fast cycle, and a revolt arrow directed toward the slow cycle have been added to the general panarchy model, highlighting the decisive role played by the intermediate level, which, in this case, corresponds to local development policies in interaction with federal resources.*

The book *Panarchy: Understanding Transformations in Human and Natural Systems* highlights learning and the capacity to learn as key factors in the resilience of human societies[16]. Human societies have a particularly strong capacity to transition from the creative destruction (Ω) phase into a renewal phase (α), drawing on accumulated capital (material resources, skills, contact networks, experience, etc.) and negotiating with behaviors which, although considered dissident during the conservation phase K, may contribute to institutional reform in passing from the Ω to the α phase (for example, see Table 12.2 in Gunderson *et al.* 2002, p. 324). However, cases of adaptation are not systematic: while human societies possess an increased innovative power, they can also be more rigid than ecological systems.

The adaptive cycle and panarchy are helpful models for representing the dynamics of social-ecological systems and resilience in a qualitative manner; ecological resilience has also led to developments allowing a more quantitative approach to the transformation of systems.

## 4.5. Alternative attractors and transitions in complex systems

### 4.5.1. *Multiple states*

Ecologists in the 1990s worked on the definition of multiple stable states, and on identifying the conditions in which social-ecological systems transition from one state to another. They notably analyzed the way in which even a minor event may trigger a transformation that then propagates, leading to a new state in the system. Researchers aimed to understand the causes that can bring complex systems to a "**tipping point**, (i.e. a catastrophic bifurcation), where a minor trigger can invoke a self-propagating shift to a contrasting state" (Scheffer *et al.* 2012, p. 344). Different terminologies were used to describe similar phenomena, meaning that this conceptual framework is not always easy to appropriate. The same types of questions come up in analyses of "alternative equilibria" (Scheffer *et al.* 1993, p. 275), "alternative stable states" (Mumby *et al.* 2007), "**alternative attractors**" (Scheffer and Carpenter 2003) or "regime shifts" (Scheffer *et al.* 2001, p. 591 and 592; Scheffer and Carpenter 2003, p. 650)[17]. Changes were generally referred to as "transitions" or "bifurcations".

A transformation is considered complete when a system leaves its basin of attraction. This situation is known as a "**catastrophic shift**" (Scheffer *et al.* 2001) and is very similar to the notion of "bifurcation", the critical point from which a radically different state may emerge (Prigogine and Stengers 1990, p. 229). In self-organization theory, a **bifurcation** corresponds to "an inflection of the trajectories of evolution toward one or other of the states of a system which are

16 See Berkes and Folke (1998), Berkes (2009) and Folke (2011).

17 The expression "alternative attractors" will be used here.

compatible with its dynamics, solutions to the mathematical system of equations”[18] (Sanders 2017, p. 35). In systems with multiple attractors, the trajectory is dependent on the initial situation. If the trajectory begins near the dividing line between two basins of attraction, even a slight variation in initial conditions may cause the system’s trajectory to tip over to the other side of the line, into the second basin, and the change will be irreversible (Sanders 1992, pp. 196–197).

There appears to be no general consensus on the vocabulary used to describe transitions. Terminology varies according to the role assigned to endogenous or exogeneous factors, or to the brutality of change (Sanders 2017, p. 39). In what follows, the terms “bifurcation“, “transition” or “regime shift“ will be used to relate to changes affecting the structure and function of a system, whatever the origin of the change (internal or external). This type of change is characterized by the emergence of a new system, with new feedback loops and a new “identity“ (Walker *et al.* 2006). The study of transitions in complex systems has many applications in different fields of research (economics, sociology, chemistry, physics, microbiology, archeology, etc.); for example, the existence of alternative attractors has been an object of study in the case of coral reefs, woodland zones, desertification, marine ecosystems and more (Scheffer *et al.* 2001, 2012).

### 4.5.2. *Alternative attractors and hysteresis*

The existence of multiple attractors can be measured empirically using the phenomenon of **hysteresis**. One commonly cited example is that of the thermostat in a heating device. The system has two states: (1) heating or (2) off. If the thermostat is set to 20°, for example, state (1) will continue until 20° is reached; above this point, it will tip into state (2), off. The temperature will then begin to fall (from point F2 in Figure 4.12). To avoid the system retriggering too frequently, a buffer is established, meaning that the heater will only start to function again at a temperature of around 19.5°, for example (point F1). There are thus two possible stable states between 19.5° and 20°: heating (state 1) or off (state 2). The difference between the forward and backward switches is known as hysteresis (Scheffer and Carpenter 2003, p. 650). In this case, the hysteresis value is 0.5 (Figure 4.11).

In the ball-in-a-valley illustration (section 4.1), hysteresis occurs if the ball is pushed into another basin of attraction (another valley) by a change in variables, then the same push occurs in the opposite direction, but is not sufficient for the ball to fall back into the initial basin of attraction (this is the case in the area between F2 and F1, before F1 is attained, in Figure 4.11):

18 *...une inflexion des trajectoires d’évolution vers l’un ou l’autre des états d’un système compatible avec sa dynamique, qui sont des solutions du système d’équations mathématiques.*

> To induce a switch back to the original valley, it is not sufficient to restore the environmental conditions present before the collapse. Instead, one needs to go back beyond another bifurcation point (F1), where the system shifts back. The difference between forward and backward switches is known as hysteresis. (Scheffer *et al.* 2003, p. 650)

In the heating example, hysteresis is the space between 20° and 19.5°. The heater will not begin to function again until the temperature reaches 19.5°. There is thus an element of irreversibility in the system trajectory. If this point (F1) is never reached, the situation may become definitively irreversible, and the system tips over into the new basin of attraction (the heater remains off). Mumby *et al.*, for example, observed this in the case of coral reefs, which have not returned to their original state in spite of a restoration of previous conditions (Mumby *et al.* 2007, pp. 98–101).

While the ball may be violently "pushed" into another valley by a significant external perturbation, transitions may also result from evolutions within the system itself. If the size of the initial basin of attraction begins to decrease, the slope separating the two valleys will shrink until a tip into the second valley becomes inevitable (Figure 4.12, profiles 3 and 4; cf. Figures 4.2 and 4.3). One sign that a system is reaching a critical point is the observation of a "critical slowing down", characterized by a slow recovery after even the smallest perturbations, indicating that the basin of attraction is shrinking (Scheffer *et al.* 2012, p. 346). This formalization of system dynamics offers the means of anticipating tipping points. According to Scheffer and Carpenter, the previous state of a system can be restored by reversing the conditions following a transition. However, simply recreating the conditions present prior to a collapse (F2) is not sufficient; another, more distant, point of bifurcation (F1) must be attained, something which Scheffer and Carpenter consider to be rare (Scheffer and Carpenter 2003, p. 650).

In the early 1990s, Scheffer *et al.* modeled the existence of at least two alternative states in shallow lakes: one in which clear water dominates, and another in which turbid water dominates. The clear water state persists because of submerged vegetation, which stabilizes the substrate and reduces the mass of phytoplankton, contributing to maintenance of its own survival conditions through a feedback loop. In cases of eutrophication, however, a lack of light prevents the development of submerged vegetation, and above a certain threshold, the system may tip over into a definitive turbid state. This state may be caused by an external perturbation (such as pollution). The existence of multiple stable states may be formalized in a graphic, showing two possible attractors for the same variable configuration value. For example, there are two possible system states for a lake:

– the first is in the basin of attraction for clear water (value at point C1 in Figure 4.11);

– the second is in the basin of attraction for turbid water (value at point C2 in Figure 4.11).

The space between the curves (BA) is a formalization of the size of the basin of attraction for clear water[19]. The higher the value corresponding to the configuration of variables, the more the basin of attraction will shrink. A perturbation (P2) may tip the system into a new state of equilibrium, while another perturbation of the same size (P1) had no effect on the system at another point in its trajectory. Scheffer *et al.* identified two ways in which a system may slide toward an alternative attractor:

1) a perturbation may cause a bifurcation in a system, if it is sufficiently large to push the system out of its initial basin of attraction (P2);

2) beyond a certain variable configuration value, the basin of attraction shrinks, and the system reaches a bifurcation point (F2); it may then tip into another basin of attraction as a result of even a very small perturbation (Scheffer *et al.* 2001, p. 592).

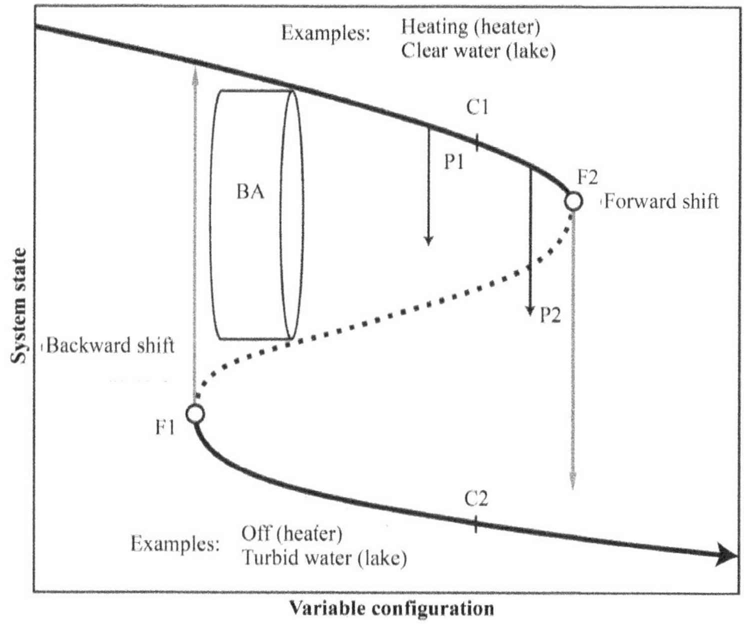


**Figure 4.11.** *Shifts between two alternative stable states in the case of a heating thermostat and of shallow lakes (adapted from Scheffer* et al. *(2001, fig. 2, 2002, fig. 8-6), by S. Robert, 2019)*

19 Following Feichtinger *et al.* (1996), who applied Holling's conceptual framework to a historical example, Scheffer *et al.* (2002) suggested a cylindrical representation of the basin of attraction.

The point at which perturbations occur within the trajectory (history) of a system thus has a decisive effect.

Scheffer *et al.* combined the curve showing alternative attractors with an image of a ball in a landscape with hills and valleys, resulting in a three-dimensional visualization that provides a clearer idea of the dynamics of these systems[20]. The situation of the system is shown at five different points on its trajectory. The first and last illustrations show a single valley: in these cases, there is only one attractor. Between these two extrema, three contexts are shown in which two attractors may coexist; this is shown by the presence of a ball in each valley (Figure 4.12, profiles 2, 3 and 4). The system thus includes alternative attractors, as there is more than one possible "stable state" for the same configuration of variables (Scheffer and Carpenter 2003, p. 651; Scheffer *et al.* 1993, p. 276, 2001, p. 593).

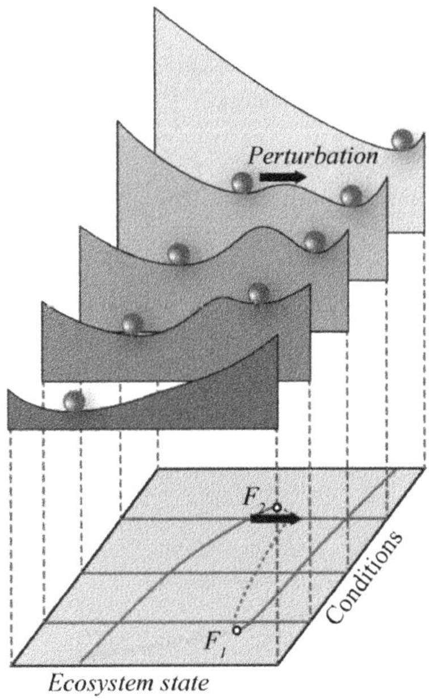


**Figure 4.12.** *"Marble-in-a-cup" representation of stability properties at five different levels of nutrient loading in shallow lakes (adapted from Scheffer* et al. *(2002, p. 204) by S. Robert, 2019). For a color version of this figure, see www.iste.co.uk/robert/resilience.zip*

20 This representation, applied to the case of shallow lakes in Scheffer *et al.* (1993, p. 275), was later generalized to other systems (Scheffer *et al.* 2001, p. 593; 2002, p. 204; Scheffer and Carpenter 2003, p. 650).

## 4.6. Conclusion

Originally used in the fields of mechanics and, later, psychology, resilience began to be adopted as part of a broader theory of adaptive complex systems from the 1970s on. It has been particularly widely used in the context of social-ecological systems, in which it is defined as:

> ...the capacity of a social-ecological system to absorb or withstand perturbations and other stressors such that the system remains within the same regime, essentially maintaining its structure and functions. It describes the degree to which the system is capable of self-organization, learning and adaptation[21].

At the same time, another definition of resilience was being developed in ecology, assimilating it to the concept of stability (calculation of the speed at which the variables characterizing a complex, stable ecosystem return to their point of equilibrium following a perturbation). In this approach, perturbations are essentially seen as exogeneous elements, endangering the stability and equilibrium of the system.

Within the conceptual framework of ecological resilience, however, change in a system is considered to be inevitable, discontinuous and unpredictable. Perturbations cannot be totally avoided, and a system's responses to these perturbations form an important part of its dynamics and of its resilience (notably insofar as it reinforces learning capacities). The theoretical framework of ecological resilience is formulated in a way which is compatible with work carried out in archeogeography from the 1990s onward; transformations in plot layouts and road networks (seen, in classical morphology, as perturbations of an initial form) are considered critical to their resilience over time (section 3.2). This is the reason ecological resilience first attracted our attention in the 2000s. Nevertheless, this framework has not been fully applied in archeogeography, a fact which will be discussed further in section 5.2.5). Concepts such as the adaptive cycle, panarchy, alternative attractors, and so on have great potential for application in this area. We must therefore consider the conditions in which the theoretical framework of ecological resistance may be transposed from the domain of ecological systems to that of the spatial systems, which form the subject of our work here.

---

21 www.resalliance.org, accessed July 30, 2019, following Holling (1973), Gunderson and Holling (2002) and Walker *et al.* (2004).

# 5

# Resilience and Spatial Systems

In this chapter, we shall consider the way in which elements of the conceptual framework of resilience can be applied to landscape and spatial systems, that is, systems "of which the component elements are spatial objects (spatial entities, portions of space which possess attributes), and of which the relationships are material or immaterial flows"[1] (Durand-Dastès and Sanders 2004). Particular attention will be paid to work in which the notions of adaptive cycles, panarchy and alternative attractors have been applied. This section is based on an analysis of published literature in both English and French in the fields of archeology and geography; we also propose some elements of bibliometric and lexical analysis. Our data were obtained manually by searching for words in the titles of books and articles, searching library catalogs and searching for certain repeated words in digitized texts. Using this approach, we were able to "map" uses of resilience and associated concepts over time and across different disciplines. Our research was carried out using the main scientific text databases that provide access to unabridged texts in digital form, and that enable users to search full texts (notably the TEL-SHS, HAL-SHS and CAIRN platforms[2]). The bulk of these resources came from PhD thesis, *habilitations à diriger les recherches* (professoral theses), texts uploaded by researchers themselves and published articles. Summary searches were also carried out using the Google Books platform and the digitized collection of old books on the French national library website (gallica.fr), but these represent only a small part of the corpus.

The first appropriations of the conceptual framework of resilience for analyzing spatial systems were made by American archeologists, then by English-speaking geographers, and finally by the French geography community. This "appropriation"

---

1 *…dont les éléments sont des objets spatiaux (des entités spatiales, des portions d'espace munies d'attributs), et dont les relations sont des flux matériels ou immatériels.*

2 Discussed in detail in section 5.2.2.

phase may be divided into two distinct periods: an "experimental" period, from the late 1970s until 2005, and a "confirmed" period, where the framework came into more general use and was the subject of significant critical debate.

## 5.1. Pre-2005: the first appropriations of resilience in archeology and geography

### 5.1.1. *Research in the English-speaking community*

From the late 1970s on, the theoretical framework of complexity and, subsequently, that of resilience, as proposed by C.S. Holling (Butzer 1980), was increasingly adopted by the English-speaking archeologists. David Yesner observes that the use of resilience increased in the 1990s, with the development of ecological theories in archeology (Yesner 2008, p. 45). Resilience was notably used in relation to themes of abandonment and desertification of territories, in connection with past periods of climate change and the degradation of the conditions of use of natural resources and land. The first of these studies concerned hunter-gatherer or early pastoral communities, who were particularly sensitive to fluctuations in the natural resources upon which they depended (Yesner *et al.* 1980; Ames 1981; Levy 1983; Yellen 1997; Gumerman and Murray 1994). Resilience also began to be used in the context of research into the collapse of highly structured societies. In 1981, Peter Timmerman published a synthesis of different models that may be used to explain the adaptation or collapse of societies as a result of climate change. He highlighted the specificities of the concept of resilience in comparison with that of vulnerability used in studies of risk: vulnerability may be assimilated to the negative reaction of a system to a dangerous event, and the degree and quality of this reaction are at least partly dependent on the resilience of the system, defined here as "the measure of a system's, or part of a system's capacity to absorb and recover from the occurrence of a hazardous event" (Timmerman 1981, p. 21).

One of the first areas in which the concept of resilience was applied in archeology relates to lost urban civilizations. Researchers such as Robert McCormick Adams or Joseph A. Tainter drew on this notion, alongside complex system or catastrophe theories[3], to explain urban transformations in ancient civilizations such as Mesopotamia, Rome, the Mayan world and the Chaco civilization (Adams 1978; Tainter 1988). In the 1990s, ecological resilience began to have a degree of influence on historical models (e.g. Feichtinger and Piccardi 1996), and first hesitant steps were taken toward its appropriation in geography. While Jonathan Sauer cited the work of C.S. Holling with respect to possible applications of ecology to biogeography (Sauer 1977, p. 324), K. S. Zimmerer barely touched on

3 Used in archeology from the 1970s on (Renfrew 1978).

resilience when discussing the possibility of integrating the "new ecology" with human geography in 1994. He cited this approach in passing, alongside a number of other approaches, and simply indicated that it could have useful applications in conservative planning (Zimmerer 1994, p. 116).

Resilience began to play a greater role in development studies in the second half of the 1990s, notably in relation to preventing coastline submersion risks. In an essay entitled *Growing with the Sea*, Helmer *et al.* (1996) put forward several ideas for increasing the resilience of the Dutch coastline by partly re-establishing the sedimentation and flooding capacities of the Rhine, Meuse and Escaut deltas. This development coincided with a tentative appropriation of the concept of resilience in development policies for vulnerable spatial systems (De Ruig 1997; Clark 1998; Klein *et al.* 1998). The mid- to late-1990s also saw the emergence of environmental concerns as a growing priority for the European Community. The Environment Agency was founded in 1993, and the principle of sustainable development was among the declared objectives of the Treaty of Amsterdam (1997). This was the context in which the *Archaeomedes* program was financed from 1992 to 1999, contributing to a diffusion of the concept of resilience in the humanities across English- and French-speaking communities (section 5.1.2.2).

In geography, an article published by the geographer Neil Adger in the journal *Progress in Human Geography* in 2000 marked an important step in the appropriation of resilience in the English-speaking community. In the article, the author discussed the relevance of resilience in risk studies, comparing it to the existing concept of vulnerability. He also showed the connections between social and ecological resilience, primarily for societies which are dependent on ecological and environmental resources for subsistence. Adger proposed a definition of social resilience as "the ability of groups or communities to cope with external stresses and disturbances as a result of social, political and environmental change (Adger 2000, p. 347)."

### 5.1.2. *The French case: simultaneous adoption in archeology and geography*

#### 5.1.2.1. *Legacies and systemogenesis: the long-term dynamics of spatial systems*

In France, the adoption of resilience built on the work of geographers who formalized the notion of spatial systems in the 1980s–1990s, considering the dynamics of these systems over long periods of time. They considered time as an essential dimension in understanding these systems:

> If a spatial object (organization, entity, etc.) is observable, it must have appeared and lasted for a sufficiently long period in order to be considered as a subject of study for geographers. Thus, in explaining this object, we must show both how it came into being and how it has survived. Geography cannot function without time[4]. (Durand-Dastès 1999, p. 23)

In the 1980s, Philippe and Pinchemel (1994) described spatial systems as sets of five sub-systems: central places, relationship networks, units of appropriation, administrative units and land uses (p. 185). R. Brunet gave a similar description based on the interactions of "four major domains of human needs and actions on the Earth"[5]: dwelling (shelter, housing), appropriation (possession, rental), exploitation (taking ownership, producing) and exchange (communication). These domains are organized, coordinated and reproduced by a fifth domain: regulation (government, maintenance, symbolic reproduction) that characterizes central places (Brunet 2001, pp. 21–31). Although each of these sub-systems has its own morphological and functional specificities, they are indissociable and interdependent, and "any transformation of one of these elements will have an impact on the others"[6] (Pinchemel and Pinchemel 1994, p. 185). From the 1980s onward, work on regional geography or demographics included the temporal dimension of spatial systems, for example, with respect to the capacity of agricultural or industrial systems to endure, notably in terms of the way they succeed, or fail, in absorbing innovations and adapting to change (Baudelle and Régnault 2004, p. 55)[7]. While this work did not make use of the conceptual framework of ecological resilience, they drew on systemic analysis approaches, considering the role of the past in present processes.

Certain geographers used the notion of **heritage, inheritance or legacy** (*heritage*) to describe the "localization process in which concrete traces left by the past are one of the causes at work in the present"[8] (Durand-Dastès 1984, p. 13). Guy Baudelle and Hervé Régnault considered heritage as something that applies to systems organized by societies, denoting a phenomenon by which "once produced, any spatial system has an effect on the economic and social system which designed it, at

---

4 *Si un "objet spatial", (une organisation, une entité), est observable c'est forcément qu'il est apparu et qu'il s'est maintenu suffisamment longtemps pour être considéré comme étudiable par les géographes. En conséquence, expliquer cet objet implique qu'on montre à la fois comment il est né et comment il a persisté. On ne peut donc pas se passer du temps pour faire de la géographie.*

5 *...quatre domaines majeurs des besoins et de l'action humaine sur la Terre.*

6 *...toute transformation d'un des éléments retentit sur les autres.*

7 For example, see Auriac (1983), Charvet (1985), Renard (1985) and Baudelle (1990).

8 *...processus de localisation dans lequel les traces concrètes laissées par le passé sont une des causes qui interviennent dans le présent.*

least in terms of the necessary difference between the spatial projects of human groups and their realization"[9] (Baudelle and Régnault 2004, p. 64). F. Durand-Dastès proposed a description of "the manifestation of the past in the present" (*manifestation du passé dans le présent*) transmitted by concrete, material phenomena in a space using other notions, such as "**inscribed memory**" (*mémoire inscrite*), the "**memory space**" (*espace mémoire*), and "**spatial**" or "**spatialized memory**" (*mémoire spatiale/spatialisée*) (Durand-Dastès 1990, p. 181; 1991). Thus, every generation has a "**received space**" (*espace reçu*), that is, a "space which possesses characteristics which are not dependent on itself"[10] (Durand-Dastès 1986, p. 116). The author considered that spatial memory was recorded through "the spatial inscription of development and transformation actions upon the earth, transmitted by the presence of concrete, material phenomena"[11]. He went on to specify that this should not be seen as a language, as the elements in question are not organized for communication purposes (Durand-Dastès 1991, p. 149). Certain events leave no trace, while others provide the initial conditions for evolutions of which the effects are still visible in the present, and which thus possess a "**localizing**" power (*localisant*) (Durand-Dastès 1991, pp. 149–151). Durand-Dastès considered that heritage was not made up of the entire past of a location, but only the part of that past which has remained effective, which has been "memorized"[12] (Durand-Dastès 1992, p. 8). The geographer defined the modes of transmission of this memory as follows:

> It is transmitted through material signs, such as places of worship, but more especially through immaterial messages transmitted from one generation to another. [...]. Past forms do not provide a direct, simple explanation for modern forms: the latter are the result of work, of choices applied to the former. [...]. Heirs are only restricted by an inheritance insofar as it is in their own interests. [...] Secondly, the role of memories is largely dependent on the strength of insertion of traces of the past into a space, and, inversely, on the ease with which they may be obliterated by the actions of new human generations[13]. (Durand-Dastès 2004, p. 22)

9 *...une fois produit, tout système spatial agit en retour sur le système économique et social qui l'a conçu, ne serait-ce qu'en raison du nécessaire décalage qui sépare les projets spatiaux des groupes humains et leur réalisation.*

10 *...espace qui a des caractères qui ne dépendent pas d'elle.*

11 *...l'inscription spatiale des actions d'aménagement et de transformation de la terre ; elle est transmise par la présence de phénomènes concrets et matériels.*

12 *...seulement la part de ce passé qui est restée efficace, qui a été "mémorisée".*

13 *Celle-ci se transmet au moyen de signes matériels, comme les lieux de culte, mais surtout à travers des messages immatériels transmis d'une génération à l'autre. [...]. Les formes*

We also encounter the idea of **inertia**, "by which an effect persists long after the cause which triggered it has disappeared"[14] (Baudelle and Régnault 2004, p. 61). For F. Durand-Dastès, inertia is also the "process by which an activity continues in a location after the initial causes have disappeared"[15]. He considered inertia, alongside heritage, as a diachronic process (Durand-Dastès 1984, p. 10 and 13). The existence of response times, reaction times or a "lag" between spatial temporalities and social or economic temporalities, highlighted by multiple authors, has been interpreted by some in terms of **hysteresis**, or a "cause-effect delay" (*retard de l'effet sur la cause*) (Baudelle and Régnault 2004, pp. 64–65). These ideas are often expressed using a type of representation commonly used in ecology, in the form of tables showing phenomena of varying spatial extent and temporal duration. Durand-Dastès, however, suggested that these different levels should be considered to be interdependent, echoing the concept of panarchy, in certain respects: "Any phenomenon of a certain order of magnitude is a set of phenomena of the order of magnitude immediately below, and one of its explanatory components is a phenomenon of the order of magnitude immediately above it"[16] (Durand-Dastès 2004, p. 20).

In the 1990s, F. Durand-Dastès worked on describing the dynamics of systems, paying particular attention to the different temporalities involved in their transformations. He proposed that "the period in which the relationships between the elements which make up a system are established"[17] should be interpreted as one of **systemogenesis** (*systémogénèse*) (Durand-Dastès 2004, p. 24). This occurs when one system takes the place of another, or during the formation of a system, that is, "the appearance of strong organizations in a situation where there were not previously any looping interactions between present elements"[18] (Durand-Dastès 2010, p. 16). Systemogenesis marks the appearance of a system within an environment that was previously only weakly organized (shown as a question mark in Figure 5.1)

*passées n'expliquent pas directement et simplement les formes actuelles : ces dernières résultent d'un travail sur les premières, d'un choix vis-à-vis des premières. [...]. Les héritiers ne sont contraints par l'héritage que dans la mesure où eux-mêmes y trouvent un intérêt. [...] En second lieu, le rôle des mémoires dépend largement de la force de l'insertion des traces du passé dans l'espace, et, partant, de la plus ou moins grande facilité avec laquelle ils peuvent être oblitérés par les actions humaines de nouvelles générations.*

14 *...ce par quoi un effet persiste bien après que la cause qui l'a déclenché a disparu.*

15 *...processus selon lequel une activité se maintient dans un lieu lorsque ses causes initiales ont disparu.*

16 *Tout phénomène d'un certain ordre de grandeur est un ensemble de phénomènes d'ordre de grandeur immédiatement inférieur et compte parmi ses composantes explicatives un phénomène d'ordre de grandeur immédiatement supérieur.*

17 *...période où se nouent les relations entre les éléments qui constitueront le système.*

18 *...l'apparition d'organisations fortes dans une situation où il n'y avait pas eu jusque-là d'interactions en boucles entre les éléments présents.*

and is characterized by "original combinations or recombinations" (*combinaisons ou des recombinaisons originales*), introducing new elements which Durand-Dastès termed ***événements***, but also *avènements*: both events and "emergences". The event is a major element in this dynamic, providing the explanation for the majority of spatial distributions. An event is defined as:

> A period of time which is relatively short in relation to the section of time in question, during which a new element appears. [...] events are localized, insofar as they occur in spaces which are relatively limited in relation to the whole area under consideration. [...] the location where a new element – a new phenomenon – emerges subsequently influences the spatial distribution of this phenomenon. [...] we move from the point (or location) to the area of distribution by combing the localized/localizing event/emergence with diffusion processes... [19]. (Durand-Dastès 2004, pp. 22–23)

These shift periods are relatively short, but they can have lasting effects, "insofar as they possess properties which render them stable or resilient, capable of maintaining themselves and of adapting"[20] (Durand-Dastès 2001, p. 9). Systemogenesis is characterized by the establishment of relationships between phenomena that were previously separate, and by the triggering of feedback loops that characterize the operation of the system (Durand-Dastès 1999, p. 11). Systemogenesis echoes the more general concept of **emergence**, used in systems and organizational analysis, which indicates that a new property or entity may be observed at a higher level of organization "as a result of interactions occurring at a lower organizational level, without that result necessarily having been desired by the actors of these interactions"[21] (Sanders 2020, p. 151). For the geographer Denise Pumain:

> [...] This idea, primarily borrowed from the life sciences, seems to add a connotation of creation, appearance or newness. In relation to the pre-existing system, emergence may consist of the creation of a

19 *Une période de temps relativement courte par rapport à la tranche de temps prise en compte, au cours de laquelle il apparaît une nouveauté. [...] les événements sont localisés, en ce sens qu'ils se produisent dans des espaces relativement limités par rapport à l'ensemble des aires envisagées. [...] le lieu où apparaît une nouveauté – un phénomène nouveau – influence ensuite la répartition spatiale de ce phénomène. [...] l'on passe du point (ou du lieu) à l'aire de répartition, en combinant l'événement/avènement localisé/localisant avec les processus de diffusion...*

20 *...dans la mesure où ils sont dotés de propriétés qui les rendent stables ou résilients, capables de se maintenir et de s'adapter.*

21 *...résultant des interactions opérées à un niveau d'organisation inférieur, sans que le résultat ait été nécessairement voulu par les acteurs de ces interactions.*

new category of elements, or the appearance of a new structure, or attributes, or in the appearance of a new structure, even a new system, or in the invention of new rules by the protagonists of a simulation game[22]. (Pumain 2003, § 19)

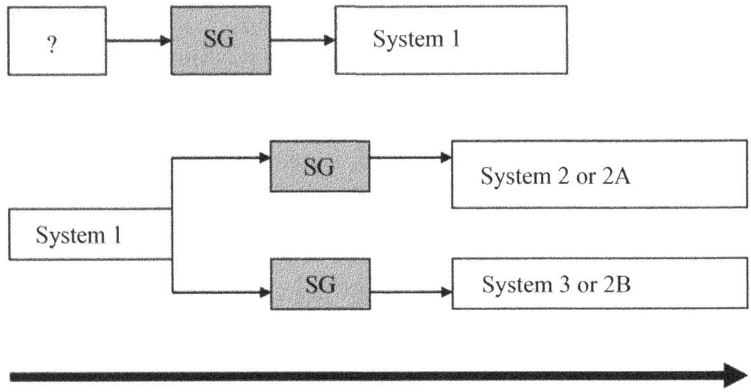


**Figure 5.1.** *Intersystemic times. Phases of systemogenesis (SG) mark the emergence of a system, or the reorganization of an existing system into a new system (source: Durand-Dastès 2018). For a color version of this figure, see www.iste.co.uk/robert/resilience.zip*

### 5.1.2.2. *Contributions of the Archaeomedes program (1992–1998)*

Beginning in the 1980s–1990s, the dynamics of spatial systems, including the temporal dimension, were a major subject of study in the field of urban geography. The theory of self-organization was applied to the study of towns and cities (Pumain *et al.* 1989), and a shared interest in the theory of dissipative structures led to increasing cooperation between geographers and archeologists (Lepetit and Pumain 1993). One interesting aspect of the application of the concept of resilience to spatial systems in France is the collaboration between geographers and archeologists in this area, notably in the wake of the *Archaeomedes* program (section 3.1.2). In 2012, looking back on their experiences in the context of the program, François Favory, Laure Nuninger and Lena Sanders noted that it acted as a "laboratory" for testing and exchanging concepts and notions between disciplines, progressively laying the foundations for a shared conceptual framework. They highlighted the heuristic contribution of the diachronic approach, which provided the tools needed to analyze

22 *Cette idée, issue principalement de l'étude du vivant, semble ajouter une connotation de création, d'apparition de la nouveauté. Par rapport au système préexistant, l'émergence peut consister dans la création d'une nouvelle catégorie d'éléments, ou d'attributs, ou dans l'apparition d'une nouvelle structure, voire d'un nouveau système, ou dans l'invention de nouvelles règles par les protagonistes d'un jeu de simulation.*

the dynamics of systems by observing their trajectories over long periods of time, from early history to the present day (Favory *et al.* 2012).

In a study of long-term settlement dynamics in the South of France, carried out within the context of *Archaeomedes*, the authors considered resilience as:

> [a] *capacity to adapt* to external change, but also to produce or adopt technical and cultural innovations, ensuring the perennity of a social group in its environment, consisting of both the natural milieu and other, competing social groups[23] (*Archaeomedes* 1998, p. 14, italics original).

Social variables with the capacity to describe continuity across highly varied periods were used to describe the conditions through which human implantations are maintained, alongside the classic functional and situational variables used in archeology and/or geography (e.g. environmental elements: relief, waterways, etc.). Variables such as the "**prior occupation value**" (*valeur de l'occupation antérieure*) or "**neighborhood legacy**" (*legs du voisinage*) were used to measure the impact of occupations of the same site over time, and of the implications of prior occupation in terms of benefits for later arrivals: the presence of road structures, the quality of land which has previously been worked, an existing drainage or irrigation network, etc. (*Archaeomedes* 1998, p. 275; Favory *et al.* 2012, p. 306). For each occupation of a site, the duration of occupation of previously occupied sites within a radius of 500 meters was counted, giving results of the form "no legacy", "less than 5 centuries", "5 to 10 centuries", etc. (*Archaeomedes* 1998, p. 257). These variables are used to distinguish between "pioneer" settlements on "pristine" ground and "opportunistic" settlements in previously occupied spaces. The study in question indicated that the foundations of the settlement structure in the South of France were already established at the time of the Roman occupation, and underwent decisive structural developments in the early Middle Ages. Later developments, in the modern and contemporary periods, were simply adaptations of pre-existing elements. The earliest settlements generally appear to have the greatest chance of survival, rapidly taking on the role of central locations where innovations tend to be adopted first (*Archaeomedes* 1998, p. 3 and 39)[24]. Researchers speak of a form of

23 ... faculté d'adaptation *à des changements externes mais aussi à la production ou à l'adoption d'innovations techniques et culturelles, qui permet d'assurer la pérennité d'un groupe social dans son environnement, lequel est constitué aussi bien par le milieu naturel que par d'autres groupes sociaux concurrents.*

24 Although the parallel was never explicitly stated, this description echoes that of pioneering and opportunistic species – r-strategists and K-strategists – in C. Holling's adaptive cycle model. The concept of a neighborhood legacy also echoes that of landesque capital (discussed in greater detail in section 6.2).

"**system memory**" (*mémoire du système*): the settlement system adapts to sociocultural changes without fundamentally altering its structure, although many fluctuations are observed (*Archaeomedes* 1998, p. 40). The system is not inert: while the overall structure is maintained, this survival is dependent on incessant transformations, passing through phases of expansion, retraction and reorganization of the habitat.

Within the *Archaeomedes* program, the geographer Christina Aschan-Leygonie studied the concept of resilience via the study of two agrarian crises in the Comtat plain area around the towns of Carpentras and Cavaillon in the Rhône Valley, France (discussed in section 6.2). The field of reference for this pioneering work (Reghezza-Zitt and Rufat 2015, p. 84) had clear connections to the American school of ecology and to work on complex systems. Aschan-Leygonie used a definition of resilience established by C.S. Holling in 1978, in which perturbations play a key role in system dynamics (Van der Leeuw and Aschan-Leygonie 2001, p. 8)[25].

Two articles were published in the wake of *Archaeomedes* to promote the concept of resilience in geographic and archeological research (Aschan-Leygonie 2000; Van der Leeuw and Aschan-Leygonie 2001). On the subject of geography, C. Aschan-Leygonie wrote, in 2000:

> One of the strong points of the notion of resilience is that it relativizes the positive character of a certain number of systemic concepts, such as stability and equilibrium. […] The main objective of this work is to explore the possibility of adapting the concept of resilience, as it is defined in ecology, to geography and specifically to the dynamics of spatial systems. This use of theoretical approaches from another discipline provides us with a better understanding of the dynamics of a spatial system in a context of perturbation[26]. (Aschan-Leygonie 2000, p. 64)

This article, published in one of the foremost French geography journals, clearly promotes the use of the ecological concept of resilience within this discipline; however, its adoption by the scientific community was far from automatic.

---

25 In the text, the authors cite Holling's 1973 text, but the definition which they give corresponds more closely to that given in 1978 (Holling 1978, p. 11).

26 *L'un des intérêts de la notion de résilience est qu'elle relativise le caractère positif d'un certain nombre de concepts systémiques, tels que la stabilité et l'équilibre. […] L'objectif principal de ce travail est d'explorer la possibilité d'adapter à la géographie, et plus particulièrement à la dynamique d'un système spatial, le concept de résilience tel qu'il est conçu en écologie. L'apport d'acquis théoriques d'une autre discipline permet, en effet, de mieux comprendre la dynamique d'un système spatial lorsqu'il est perturbé.*

### 5.1.3. *Geography: a limited adoption of the concept of resilience*

In 2000, the geographer Bertrand Elissalde noted that the concept of resilience, promoted by C. Aschan-Leygonie based on the work of C.S. Holling, was an innovative approach, one which "reconciles durability and evolution", "moving beyond the aporia of inertia and 'immobile time' applied to space". It notably shows that "the stability of systems is not necessarily synonymous with stationarity"[27] (Elissalde 2000, p. 232). The concept of resilience also features in *Échelles et temporalités en géographie* (Scales and Temporalities in Geography), a handbook published by Guy Baudelle and Hervé Régnault in 2004. Baudelle and Régnault discussed Aschan-Leygonie's work alongside the use of resilience in geomorphology, particularly with regard to the elasticity of materials (Baudelle and Régnault 2004, pp. 54–56). In 2005, however, the geographer Géraldine Djament expressed a preference for the term *reproduction spatiale*, spatial reproduction, reflecting the idea of "social reproduction" used in sociology. Djament studied the continuity of the centrality of the Roman pontifical system from ancient times to the present day, showing that the permanence of Roman urban centrality was only possible because of constant metamorphoses and regular changes in scale, from a medieval system centered on the city of Rome, then from Europe to the modern world system (Djament 2005; Djament-Tran 2011).

In practice, resilience does not seem to have been truly adopted by the geography community until the mid-2000s. While the term was included in Brunet *et al.*'s *Dictionnaire de la géographie*, published in 1993, it did not feature in another *Dictionnaire de la géographie* published in 2003 by Jacques Lévy and Michel Lussault (second edition 2013), or in the one published by Yves Lacoste in 2003, or in the second, updated edition of Pierre Georges and F. Verger's *Dictionnaire de la géographie* in 2006 (Brunet *et al.* 1993, p. 432; Lévy and Lussault, 2003; Lacoste 2003; Georges and Verger 2006). A search for the word "resilience" in four generalist French-language geography journals[28] shows a first significant increase in instances in 2012 (Figure 5.2). For the whole period from 1891 to 2007, only around 15 occurrences are listed. The term features most heavily in *L'Espace géographique*

27 *...concilier durabilité et évolution [...] dépasser les apories de l'inertie et du "temps immobile" appliqués à l'espace [...] que la stabilité des systèmes n'est pas forcément synonyme de stationnarité.*

28 *Annales de géographie*, established in 1891 by Paul Vidal de La Blache, *L'Information géographique*, established in 1936 by André Cholley, and *L'Espace géographique*, established in 1972 by Roger Brunet, are generalist geography journals with a particular focus on emerging concepts in the field. *Espaces et Sociétés*, established in 1970, focuses on urban development and architecture.

and *Annales de géographie*[29]. From our analysis, we see that the concept of resilience was only really adopted in the field of geography in France in 2012, when it appears essentially as the subject of critical debate (discussed in greater detail in section 5.2). The year with the highest rate of occurrence (2012) corresponds to the publication of *L'Espace géographique* no. 41, which featured a thematic section on the connections between archeology and geography, directed by the author of this book. Five of the nine recorded instances of the term for 2012 were found in this specific journal (Robert 2012a).

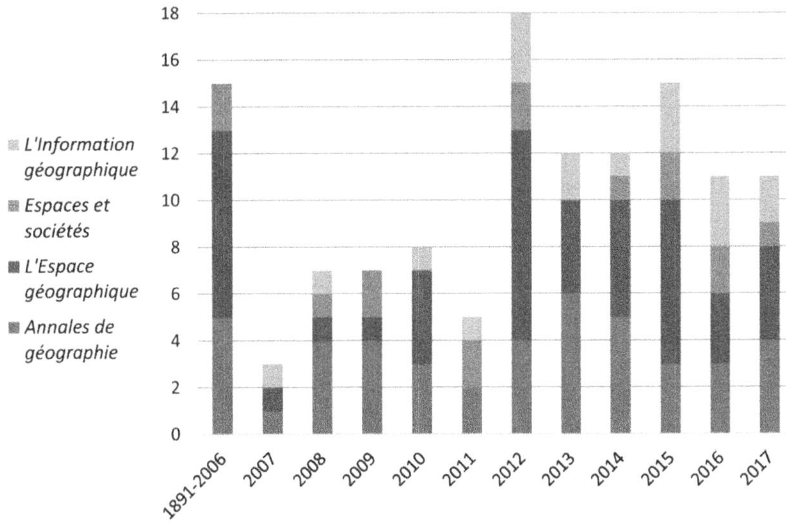


**Figure 5.2.** *Occurrences of the term "resilience" in four main French geography journals between 1891 and 2017 (Robert 2018; data source: www.cairn.info, consulted June 13, 2018). For a color version of this figure, see www.iste.co.uk/robert/resilience.zip*

### 5.1.4. *Archeology: an adoption of resilience mediated by archeogeography*

While the concept of resilience was already being used in the English-speaking archeology world from the 1970s–1980s, it was relatively slow to be adopted in the French-speaking field. Given the limited availability of the main archeological journals online, it has not been possible to analyze occurrences of the term as we did for geographical journals; instead, we have simply counted the number of articles per year using the term "resilience" in the journal *Les Nouvelles de l'archéologie*, for

29 Analysis carried out using search filters in the CAIRN database: www.cairn.info, accessed June 13, 2018.

which archives are available for the years 1979–2018 (Figure 5.3). *Les Nouvelles de l'archéologie* is a generalist publication, established in 1979, as "a place for debate and a reflection of evolutions in the discipline"[30]. The word "resilience" appears for the first time in 2009, and has only been used sporadically from 2009 to the present day. The highest occurrences correspond to special dossiers within the journal. The first two cases relate to archeogeography, including a dossier on the subject of roads and pathways which we produced in collaboration with N. Verdier, *Pour une recherche sur les routes, voies et réseaux* (Robert and Verdier 2009), and the dossier *L'archéogéographie : Un état des lieux et de leurs dynamiques* directed by M. Watteaux in 2011 (Watteaux 2011). The highest number of occurrences, in 2016, were found in a dossier named *Environnement et sociétés* (Giligny and Tsirtsoni 2016).

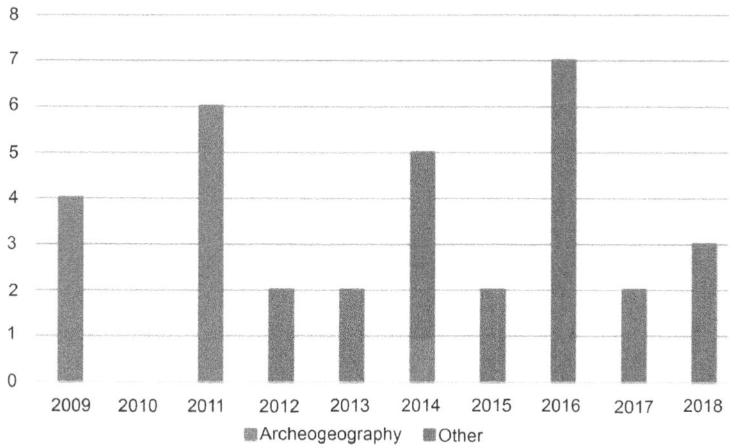


**Figure 5.3.** *Number of articles using the term "resilience" in Les Nouvelles de l'archéologie between 1979 and 2018 (Robert 2019; data source: https://journals.openedition.org/nda/, consulted March 15, 2019). For a color version of this figure, see www.iste.co.uk/robert/resilience.zip*

In 2012, a search for the word "resilience" or "resilience" in the FRANTIQ catalog[31], which includes the catalogs of all of the main archeological research libraries, returned only five references, all in English. In 2018, the same search returned 106 results, but only 20 in French; 14 of these were connected to the publication of the *Dynamique et résilience des réseaux routiers en région*

30 *Les Nouvelles de l'archéologie*. http://journals.openedition.org/nda, accessed June 26, 2019. *...à la fois un lieu de débat et le reflet des évolutions de la discipline.*

31 FRANTIQ: *Fédération et ressources sur l'Antiquité*, Federation and Resources for Antiquity. https://www.frantiq.fr/fr/missions, accessed May 10, 2018.

*Île-de-France* program, which we directed in collaboration with N. Verdier from 2006 to 2014. The adoption of resilience in the early 2000s was essentially limited to the domain of archeogeography; the dynamic character of the concept, along with the role assigned to perturbations, attracted the attention of researchers in the field. Writing in 2003, C. Marchand noted that:

> It is, however, essential that we do not make resilience into a uniquely adaptive property, a means of absorbing external perturbations, that we do not consider it simply as the capacity to survive in spite of these perturbations. To do so would be to forget that these perturbations drive dynamics through continual change, making durability possible. It is thanks to them that the system will be able to reproduce itself[32]. (Marchand 2003, p. 105)

Archeogeographic analysis thus appears to be particularly in phase with the definition of resilience given by Holling in 1978, as "a property that allows a system to absorb and *utilize* (or even benefit from) change" (Holling 1978, p. 11). The study of road networks or parcel layouts over long periods has shown that the transmission of these elements over time is dependent on the integration of transformations, rather than the maintenance of original forms (section 3.5.3).

In the first half of the 2000s, resilience was also used in the context of collaborative research by archeologists and geomorphologists on the subject of river systems (Burnouf 2009). The book *Temps et espaces des crises de l'environnement* (Time and Spaces of Environmental Crises), published in 2006 (Beck *et al.* 2006), is one of the few French-language publications to present Holling's adaptive cycle and panarchy concepts, both featuring in a chapter by S. Van der Leeuw (Van der Leeuw 2006). Thus, use of resilience in French archeology in the early 2000s remained limited to a relatively small group of researchers, notably centering on the *Archaeomedes* program and on the *Archéologies environnementales* study group, UMR 7041 (Paris, Nanterre).

Our analysis of the adoption and spread of the concept of resilience, used to express the maintenance of settlement or landscape structures, in archeology is based on a quantification of occurrences of the term in archeological theses and publications. This quantification is compared to occurrences of other terms (persistence, permanence and transmission). Our corpus is made up of 33 references, published between 1996 and 2017: 11 books, 13 theses, five articles in thematic journals, two

32 *Il faut prendre garde cependant à ne pas faire de la résilience une propriété uniquement adaptative, d'absorption des perturbations externes, à ne pas la considérer seulement comme la capacité de perdurer malgré ces perturbations. Ce serait en effet oublier que ce sont les perturbations qui constituent le moteur de la dynamique par des changements continuels et qui rendent possible la durabilité. C'est grâce à celles-ci que le système va pouvoir se reproduire.*

HDR professorial theses, a thematic journal issue and a program report[33] (Table 5.1, Figure 5.4). A distinction is made between three areas of research:

1) spatial analysis, which concerns settlement networks, in terms of their distribution in relation to, or independent of, circulation networks;

2) morphology, which concerns the permanence of urban or rural forms, in the way traditionally used in history and art history;

3) archeogeography, a development of morphological analysis that features a strong geographical influence, from the 1990s onward.

Instances of the term "resilience" are compared with uses of the term "self-organization" (*auto-organisation*) to verify whether the use of the concept of resilience ran parallel to an increasing interest in complex systems. We carried out systematic searches for these different terms in texts for which we were able to obtain a full digital version. Our analysis is only partial, due to the limitations of access to these texts, but it does provide some indications concerning the adoption of resilience and on the terms used to describe the maintenance of spatial systems in the 1990s–2000s.

The term "permanence" appears most frequently in our analysis (in 30 out of 33 references). This term is traditionally used in history and geography (e.g. in the work of F. Braudel), and was the dominant term in archeogeography in this respect in the 1990s. The term "resilience" was introduced in the context of the *Archaeomedes* program from 1998 onward, associated with the theory of self-organization. It became more common in archeogeography from the early 2000s on (Chouquer 2000; Marchand 2003; Robert 2003b). In 2006, we chose to use the term in the title of a collective research program, *Dynamique et résilience des réseaux routiers en région Île-de-France* (*Dynarif*) (Dynamics and Resilience of Road Networks in the Ile-de-France Region). In spatial archeology and morphology, however, the term was far less common, although Pierre Garmy did highlight its potential for the study of settlement structures in 2009:

> In this respect, the generalization of the concept of resilience, adapted from the physical sciences, notably by ecologists (Holling 1973), which has permeated the humanities, notably geography, to some extent (Aschan-Leygonie 2000), would be beneficial for the archeological study of settlement structures[34]. (Garmy 2009, p. 154)

33 This choice was partly determined by access to digital versions of full texts.

34 *À cet égard, le concept de résilience, adapté des sciences physiques par les écologues notamment (Holling 1973) mais qui tend à se diffuser à l'ensemble des sciences humaines et en particulier en géographie (Aschan-Leygonie 2000), trouverait de manière profitable une généralisation de son emploi dans l'étude archéologique des structures de peuplement.*

| Reference | Type | Year | School | Persistance | Permanence | Transmission | Resilience | Self-organization |
|---|---|---|---|---|---|---|---|---|
| Chouquer (ed.) 1996a | Book | 1996 | archeogeography | 0 | 7 | 0 | 0 | 0 |
| Chouquer (ed.) 1996b | Book | 1996 | archeogeography | 1 | 12 | 3 | 0 | 0 |
| Chouquer (ed.) 1997 | Book | 1997 | archeogeography | 1 | 10 | 0 | 0 | 0 |
| Van der Leeuw (ed.) 1998 | Book | 1998 | spatial archeology | 0 | 0 | 2 | 139 | 4 |
| *Archaeomedes* 1998 | Book | 1998 | spatial archeology | 4 | 4 | 0 | 7 | 18 |
| Jung 1999 | Thesis | 1999 | archeogeography | 0 | 18 | 0 | 0 | 0 |
| Marchand 2000 | Thesis | 2000 | archeogeography | 0 | 4 | **15** | 0 | 15 |
| Chouquer 2000 | Book | 2000 | archeogeography | 0 | 16 | 2 | 5 | 2 |
| Chevallier 2000 | Book | 2000 | morphology | 2 | 5 | 3 | 0 | 0 |
| Nuninger 2002 | Thesis | 2002 | spatial archeology | 0 | 2 | 0 | 0 | 0 |
| Robert 2003a, T2 | Thesis | 2003 | archeogeography | 8 | 10 | 25 | **35** | 20 |
| Chouquer (ed.) 2003 | Journal | 2003 | archeogeography | 13 | 17 | 51 | 13 | 24 |
| Abbé 2006 | Book | 2006 | morphology | 1 | 2 | 3 | 0 | 0 |
| Gandini 2006 | Thesis | 2006 | spatial archeology | 2 | 7 | 1 | 1 | 0 |
| Chouquer 2007a | Book | 2007 | archeogeography | 4 | 10 | 37 | 13 | 20 |
| Poirier 2007 | Thesis | 2007 | spatial archeology | 1 | 2 | 1 | 1 | 1 |
| Favory *et al.* 2008 | Report | 2008 | spatial archeology | 1 | 1 | 0 | 0 | 0 |

| Lefebvre 2008 | Thesis | 2008 | morphology | 12 | 11 | 0 | 0 | 3 |
|---|---|---|---|---|---|---|---|---|
| Garmy 2009 | HDR | 2009 | spatial archeology | 3 | 12 | 13 | 4 | 11 |
| Watteaux 2009 | Thesis | 2009 | archeogeography | 13 | 18 | 125 | 75 | 2 |
| Fovet 2010 | Thesis | 2010 | spatial archeology | 2 | 1 | 0 | 1 | 0 |
| Brigand 2010 | Thesis | 2010 | archeogeography | 2 | 1 | 85 | 11 | 4 |
| Robert (ed.) 2011 | Book | 2011 | archeogeography | 1 | 3 | 48 | 2 | 6 |
| Robert 2012 | Article | 2012 | archeogeography | 1 | 1 | 19 | 11 | 4 |
| Favory *et al.*2012 | Article | 2012 | spatial archeology | 0 | 2 | 0 | 1 | 0 |
| Noizet *et al.* 2013 | Article | 2013 | archeogeography | 0 | 0 | 13 | **18** | 0 |
| Robert and Verdier (eds) 2014 | Book | 2014 | archeogeography | 0 | 0 | 17 | **34** | 0 |
| Rodier 2016 | HDR | 2016 | spatial archeology | 0 | 5 | 4 | 0 | 3 |
| Burnouf and Gillot 2016 | Article | 2016 | archeogeography | 1 | 3 | 5 | 17 | 2 |
| Cavanna 2016 | Thesis | 2016 | archeogeography | 1 | 4 | 24 | 37 | 2 |
| Fusco 2016 | Thesis | 2016 | spatial archeology | 3 | 0 | 4 | [illegible] | 0 |
| Hermenault 2017 | Thesis | 2017 | archeogeography | 3 | 8 | 25 | 33 | 2 |
| Robert 2017 | Article | 2017 | archeogeography | 0 | 0 | 0 | 18 | 3 |
| **Total** | | | | 80 | 189 | 525 | 477 | 146 |
| **Frequency** | | | | 23 | 30 | 23 | 22 | 20 |

**Table 5.1.** *Terms used to express the maintenance of forms in a selection of French-language publications (except Van der Leeuw 1998, in English), on the analysis of road networks, plot layouts and settlement systems in archeology and archeogeography between 1996 and 2017, giving the number of occurrences of each term in the document. Shaded squares show the most common term in each text. Bold text shows turning points discussed in greater detail below (Robert 2019)*

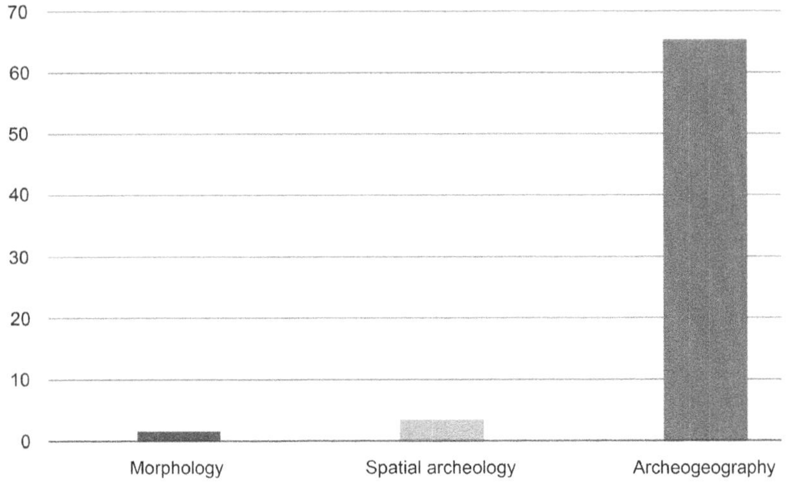


**Figure 5.4.** *Occurrences of the term "resilience" as a percentage of total analyzed terms in a selection of French publications on the subject of road networks, parcel layouts and settlement systems in archeology and archeogeography between 1996 and 2017 (Robert 2019; source: Table 5.1)*[35]*. For a color version of this figure, see www.iste.co.uk/robert/resilience.zip*

The first appearance of the term "resilience" in book, article and seminar titles occurred in 2012, the same year it overtook "transmission" in archeogeography publications. Much of this corresponds to our own efforts to develop the application of the conceptual framework of resilience in archeogeography. In 2012, the term "resilience" was included in the title of my own chair at the EHESS, and the concept is one of the major elements of our research and teaching program, *Fabrique, transmission et résilience des formes: archéogéographie* (Creation, Transmission and Resilience of Forms: Archeogeography)[36]. In 2013 and 2014, we chose to include the term in the title of publications on the Seine paleomeander and of the

35 For spatial archeology, the only counted instances are in the French publication resulting from the *Archaeomedes* program (*Archaeomedes* 1998). The English publication (Van der Leeuw 1998) was excluded in order to focus on the adoption of the term in French.

36 Titles of annual research seminars on resilience have included *Résilience de longue durée et réseaux informels de circulation* (long-term resilience and informal circulation networks, 2013–2014); *Le paysage comme trace : un laboratoire pour le concept de résilience* (landscape as trace: a laboratory for the concept of resilience, 2017–2019); and *Le paysage comme système socio-écologique résilient* (the landscape as a resilient social-ecological system, 2019–2020). For details, see the annual reports, course reports and conference papers of the Ehess for the corresponding years.

*Dynarif* program (Noizet *et al.* 2013; Robert and Verdier 2014). In 2016, resilience was selected as the theme for the *Laboratoire d'excellence* summer school, *Dynamiques territoriales et spatiales* (Labex DynamiTe)[37] and, in 2018, I ran a session on the use of the conceptual framework of resilience in landscape archeology and archeogeography as part of the 18th World Congress of the International Union for Prehistoric and Protohistoric Sciences in Paris[38]. Since 2015, the term "resilience" has overtaken "transmission" in theses in the field of archeogeography. However, it is most often used in its "standard" sense, to describe the reproduction of parcel layouts and road networks over time, without truly drawing on elements of the conceptual framework of ecological resilience (basins of attraction, adaptive cycles, panarchy, alternative attractors, etc.) (section 5.2.6).

This re-mobilization of the concept of resilience in archeogeography has taken place against a very different background to that encountered in the early 2000s. It is important to take account of debates concerning the use of the term, in the wake of its wholesale appropriation by public authorities and throughout the humanities in the English-speaking world from the late 2000s on.

## 5.2. New developments and critical approaches after 2005

### 5.2.1. *Adoption of the notion of resilience by public authorities and inflation of the term*

The mid-2000s and early 2010s were marked by a rapid upsurge in uses of the word "resilience" in the titles of publications, a fact illustrated by the growing number of documents cataloged by the Bibliothèque nationale de France with "resilience" or "resilience" in their titles. Figure 5.5 shows an "explosion" in uses of the term from the 2000s on, becoming even more striking in the 2010s. The first recorded title is from 1953; in 1999, there were still only 23 documents with "resilience" or "resilience" in their titles. The 474 occurrences recorded between 2000 and 2018 are thus even more striking. Up until the 1990s, the vast majority references in the catalog related to the fields of physics or metals engineering. A few

37 Labex DynamiTe Summer School, *ANR*, managed by the University of Paris 1 Panthéon-Sorbonne: *Du risque à la résilience, du passé au présent: perspectives critiques et approches comparatives* (From risk to resilience, past to present: critical perspectives and comparative approaches), coordinated by Sandrine Robert, Frank Lavigne and Véronique Dupont, University of Catania, July 6–12, 2016. http://labex-dynamite.com/fr/evenements-du-labex/ecoles-dete/ecole-dete-2016.

38 *Résilience et paysage* dossier in the digital journal *Archéologie Société Environnement* (vol. 21-2, no. 1, 2021), ISTE Openscience: https://www.openscience.fr/Archeologie-societe-et-environnement.

English-language references from the field of anthropology made their appearance in the 1980s. From the 2000s to the 2010s, French-language results were almost exclusively found in the domains of psychology and the educational sciences; Vincent Soulié and Boris Cyrulnik spoke of resilience as a "new psychological concept" (*nouveau concept psychologique*: Soulié and Cyrulnik 2000). When the *Que sais-je ?* collection[39] brought out a volume on resilience in 2007, it related exclusively to the field of psychology (Tisseron 2007). Titles mentioning resilience in the context of sustainable development and risk management began to appear in the Bibliothèque Nationale catalog in the first decade of the 2010s, but all of these publications were in English. A few French-language works relating to resilience in the context of business management or law were added to the catalog from 2005. The first documents on the subject of climate crises or territorial studies with the word "résilience" in their title were added in 2013.

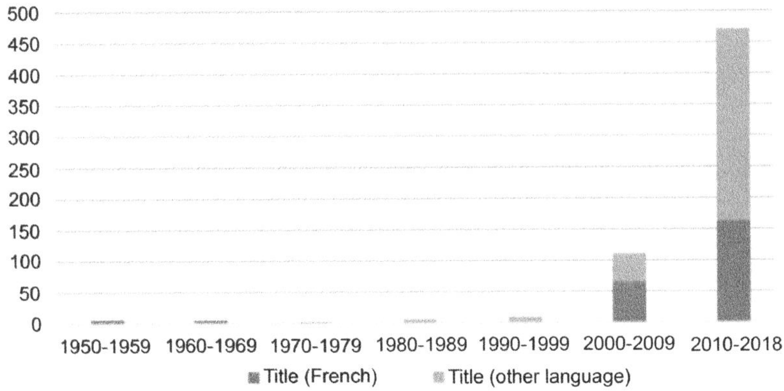


**Figure 5.5.** *Number of documents added to the Bibliothèque nationale de France (BNF) catalog containing the word "resilience" or "resilience" in their title (Robert, 2019; source: https://catalogue.bnf.fr, accessed January 9, 2019). For a color version of this figure, see www.iste.co.uk/robert/resilience.zip*

This inflation in the French context reflects the general trend observed for English-language publications. A similar analysis of the titles of printed or digital texts held by the Library of Congress shows four times more instances of the term *resilience* than in the BNF catalog (2,523 occurrences, in 2018, compared to 619). There are virtually no references predating the 1970s; subsequently, the term appeared in titles on the subject of metallurgy, then ecology, economics or national security. In English, as in French, a major upsurge is observed in the 2000s,

39 Author's note: *Que Sais-Je ?* is a famous French collection of short introductory works on various, mostly academic, themes. As of 2021, 1,525 volumes are currently available, out of a total of over 3,800.

particularly after 2010 (Figure 5.6). This observation corresponds to the findings of a search using the Web of Science portal[40] in December 2015, cited by C. Folke, indicating a 25-fold increase in scientific publications on the subject of resilience in an environmental context over a 15-year period (Folke 2016, p. 3).

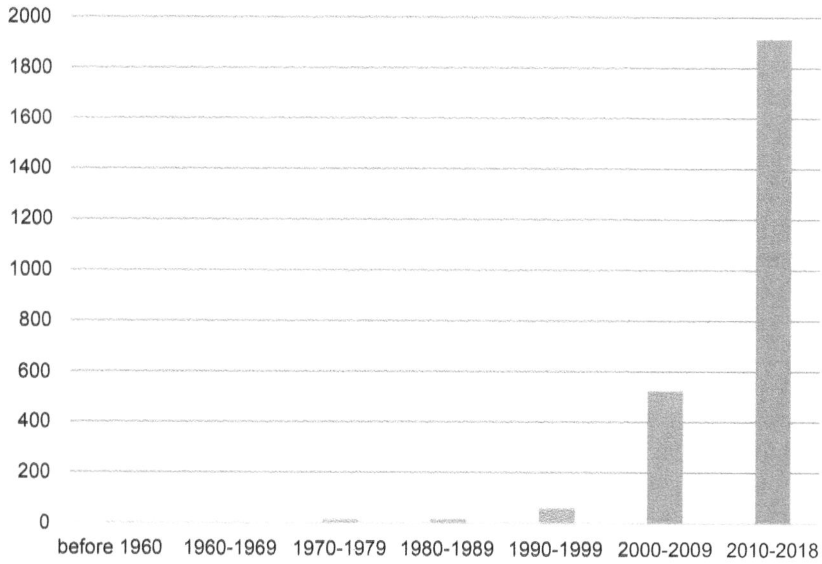


**Figure 5.6.** *Number of documents recorded in the US Library of Congress catalog featuring the word "résilience" or "resilience" in their title (Robert 2019; source https://catalog.loc.gov/, accessed January 9, 2019). For a color version of this figure, see www.iste.co.uk/robert/resilience.zip*

The upsurge in uses of the word "resilience" in scientific literature should be interpreted in conjunction with the adoption of notions of resilience in public policy and by private foundations, which invested in research on this theme from 2005 onwards. In 2005, the UN used the term in its framework for action for the period 2005–2015, "Building the resilience of nations and communities to disasters", presented at the World Conference on Catastrophe Reduction in Kobe (Hyogo, Japan)[41]; this marked an important step in terms of institutional adoption. The

40 The Web of Science portal is an index of peer-reviewed scientific journals, mostly in English and focused on the American research community (https://clarivate.com/products/web-of-science/, accessed October 1, 2018).

41 "Hyogo Framework for Action 2005–2015: Building the resilience of nations and communities to disasters – UNISDR". https://www.unisdr.org/we/inform/publications/1037, accessed February 20, 2018.

framework echoed overarching concerns in terms of risk reduction and crisis management, at a time when Western societies were searching for responses to a certain number of threats (terrorism, natural disasters, increased vulnerability of certain categories of the population, etc.). The terrorist attacks of September 11, 2001, and the effects of Hurricane Katrina in 2005 seem to have had a particularly strong impact in terms of interest in the concept of resilience from the mid-2000s on (Vale and Campanella 2005).

Over the following years, several international institutions placed resilience at the heart of their programs[42]. The report of the United Nations Secretary-General's High-Level Panel on Global Sustainability, presented at the United Nations Conference on Sustainable Development, Rio+20, in 2012, was entitled *Resilient People, Resilient Planet: A Future Worth Choosing*[43]. In 2013, the theme of the World Economic Forum at Davos was "Resilient Dynamism", in connection with global risks[44]. That same year, the Rockefeller Foundation created the "100 Resilient Cities" program, aimed at "helping cities around the world become more resilient to physical, social, and economic shocks and stresses"[45]. The 100 resilient cities formed part of a Global Resilience Partnership, directed by the Rockefeller Foundation in association with the United States Agency for International Development (USAID) and with the Swedish International Development Cooperation Agency (Sida)[46]. In the 2010s, associations of private entities, such as insurance companies or multinational firms, also financed their own programs to strengthen the adaptive capacity of the communities with which they interacted, for example with regard to flood risks, food shortages, energy crises, etc. (Folke 2016, p. 2). From the second half of the 2000s, resilience also appeared as a key element in the governing principles of certain alternative development movements, such as the Transition Towns Movement, founded in 2005, and the development of permaculture (Hopkins 2008).

This growing interest in resilience found an echo in the humanities and social sciences, and should be considered in conjunction with debates and reflections on the subject of Global Change. In 2006, the mathematician Marco A. Janssen and the

42 For a detailed analysis of the use of resilience by international, European and French authorities, see Bourcart (2015).

43 The group presented 56 recommendations for "choosing resilience" and for sustainable development. http://www.un.org/press/fr/2012/AG11212.doc.htm, consulted 4th June 2020.

44 World Economic Forum Annual Meeting 2013. World Economic Forum. https://www.weforum.org/reports/world-economic-forum-annual-meeting-2013/, accessed June 4, 2020.

45 http://www.100resilientcities.org/, accessed November 13, 2017.

46 "Global Resilience Partnership: An Overview". The Rockefeller Foundation. https://www.rockefellerfoundation.org/report/global-resilience-partnership-overview/, accessed June 4, 2020.

economist Elinor Ostrom directed a special edition of the *Global Environmental Change* journal, entitled "Resilience, Vulnerability and Adaptation. A cross-cutting theme of the International Human Dimension Programme [IHDP] on Global Environmental Change" (Janssen and Ostrom 2006). This publication appeared in the wake of an IHDP seminar held in February 2005 at Arizona State University, bringing together experts from across the humanities to work on the connections between the concepts of vulnerability, resilience and adaptation to climate change, and to develop a specification grid for their use. The French geographer Pierre-Olivier Garcia considered this publication as a *mise en paradigme* (establishing a paradigm) for resilience in the late 2000s, reconfiguring the discourse on adaptation to climate change within the scientific community (Garcia 2015).

The adoption of the notion of resilience by international authorities, management agencies and the scientific community *en masse* was clearly a reflection of the times; the upsurge in interest in the late 2000s and particularly the 2010s earned it the status of a "buzzword"[47]. This led to debates concerning the use of resilience in operational contexts, but also in its applications within the humanities from the second half of the 2000s onward.

### 5.2.2. *The adoption of resilience in the geography of risks*

The concept of resilience was adopted by the English-speaking geography community in the mid-2000s, notably in the domain of risk management and in relation to questions of global change (Brown 2014). Activity was particularly strong in the area of urban resilience. In the wake of the terrorist attacks of September 11, 2001, Lawrence J. Vale and Thomas J. Campanella of the Department of Urban Studies and Planning at MIT organized a conference on the theme of "The Resilient City, Urban Trauma, Recovery and Remembrance" (Vale and Campanella 2005). Participants examined the post-catastrophic resilience capacities of different cities (such as Berlin after 1945, Mexico City following the 1985 earthquake, etc.). Outside the field of catastrophe studies, urban resilience also drew on research on the systems of production in relation to towns and cities, themselves viewed as ecosystems (Folke 1997; Bengtsson *et al.* 2003; Newman and Jennings 2008). In urban development, Sara Meerow, Joshua P. Newell and Melissa Stults analyzed the network of co-citations involving the terms *urban resilience* and *resilient cities* in publications recorded in the Elsevier's Scopus and Thompson Reuters Web of Science databases from 1973 to 2013. They showed that the intellectual structure used in publications on urban development is strongly marked

47 Walsh B., Adapt or Die: Why the Environmental Buzzword of 2013 Will Be Resilience, *Time* 2013, accessed April 13, 2018; see also Comfort *et al.* (2010).

by ecological resilience. Several publications, including C.S. Holling's 1973 article, are seen as highly centralizing nodes in the co-citation graph (Meerow *et al.* 2016).

In France, use of the concept of resilience became more widespread in geography after 2005. This development occurred not in the analysis of settlement systems or agrarian systems, where resilience was first introduced, but in risk analysis, notably following the publication of an article by André Dauphiné and Damienne Provitolo in 2017, in which resilience was presented as "an operational concept for risk management"[48] (Dauphiné and Provitolo 2007). Resilience began to appear at the heart of publications and theses in geography at the end of the 2000s (e.g. Maret and Cadoul 2008; Pierdet 2009; Hernandez 2010; Lavigne and de Belizal 2010; Cotelle 2014).

The *Centre pour la communication scientifique directe* (Center for Direct Scientific Communication, CCSD) runs a website, TEL (thèses-en-ligne), to which doctoral theses presented in France were uploaded on a voluntary basis from 2001 to 2015, then systematically from 2016[49]. The uploading of HDR professorial theses is still voluntary. Between 1997 and 2018, there were 162 archived documents across all disciplines featuring the keywords "resilience" or "résilience" (154 doctoral theses and eight HDRs) (Figure 5.7). The highest number related to the humanities and social sciences (86 occurrences), followed by computer science (28), engineering sciences (27), the life sciences (17) and environmental sciences (14)[50]. The first occurrence in the humanities and social sciences, in 2003, corresponds to the author's own thesis (Robert 2003b)[51]. Four of the eight self-uploaded HDRs were in the humanities and social sciences. The first, from 2011, was in the field of architecture and spatial development (Serre 2011), while the remainder were in the field of geography (Reghezza-Zitt 2015; Fusco 2018; Le Blanc 2018). As in our earlier analyses, we see a clear upsurge in uses of the term in the 2010s[52]. For example, the first HDR to cite "resilience" as a keyword was uploaded in 2011, and the term was first used in the title of an uploaded HDR (in the computer sciences) in 2013.

48 *...un concept opératoire pour la gestion des risques.*

49 The CCSD was created by the CNRS in 2000. The TEL (Thèses-en-Ligne) database, created in 2001, was originally used by researchers to upload doctoral and professoral theses (HDR) on a voluntary basis. Certifying establishments use the STAR (*Signalement des thèses et archivage*) application to make theses available through TEL (www.ccsd.cnrs.fr, consulted June 25, 2019). The use of STAR was made compulsory for all establishments delivering doctoral qualifications by a law of May 25, 2016 (www.abes.fr, accessed June 25, 2019).

50 Theses may be listed in multiple fields.

51 C. Aschan-Leygonie's thesis (1998) was not uploaded to the database.

52 The peak for the year 2016 may also be a result of the fact that the uploading of theses to TEL became compulsory after this date.

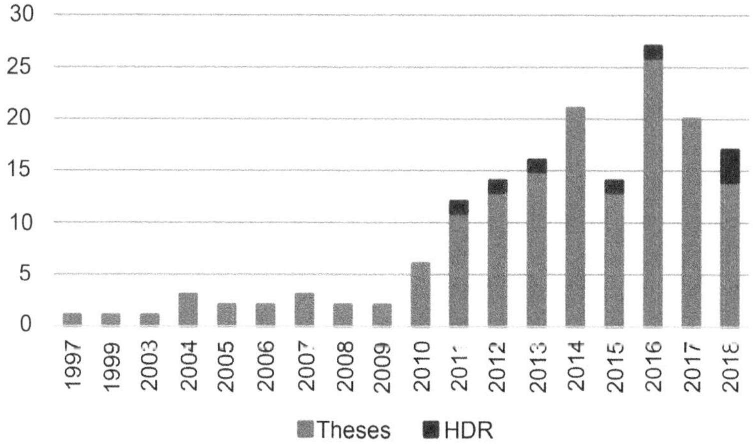


**Figure 5.7.** *Occurrences of the word "résilience" or "resilience" in the keywords of theses and HDRs uploaded to the CCSD's TEL (Thèses-en-Ligne) platform between 2001 and 2018 across all disciplines. The first occurrence in the humanities listed on TEL dates from 2003 (Robert 2003b). The first uses of the term in HDRs appear from 2011 on (Robert 2019; source: https://tel.archives-ouvertes.fr/, accessed June 25, 2019). For a color version of this figure, see www.iste.co.uk/robert/resilience.zip*

This analysis was compared to occurrences of the word "resilience"[53] in the CAIRN Internet portal. Founded in 2005, CAIRN holds the catalogs of over 150 publishers in the humanities, along with the full texts of journals and works published since 2001. Searching the full texts available on the platform, we identified 6,970 occurrences of the word "resilience" across 17 different disciplinary fields[54]. Geography accounted for 1,149 occurrences (16.5%), putting it in fifth place, after psychology, economics, political science and sociology. The rankings for archeology and archeogeography cannot be defined precisely due to the absence of a dedicated category; however, these fields are generally categorized as history, in sixth place (out of 17) with 7.8% of occurrences (Figure 5.8).

Between 2009 and 2012, some of the work carried out on this subject in geography was examined in the context of the *Résilience Urbaine* (Urban Resilience) seminar, held at the *École normale supérieure* by Géraldine Djament and Magali Reghezza-Zit. This results in the publication of two works on urban resilience, which examine the critical dimension of the notion of resilience itself

53 The CAIRN search engine does not differentiate between accented and non-accented letters, hence the single keyword.

54 Disciplinary fields are non-exclusive.

(Djament-Tran and Reghezza-Zitt 2012; Reghezza-Zitt and Rufat 2015)[55]. The increasing adoption of resilience in the social sciences, notably in the geography of risks, gave rise to significant debate in both the English and French-speaking research communities.

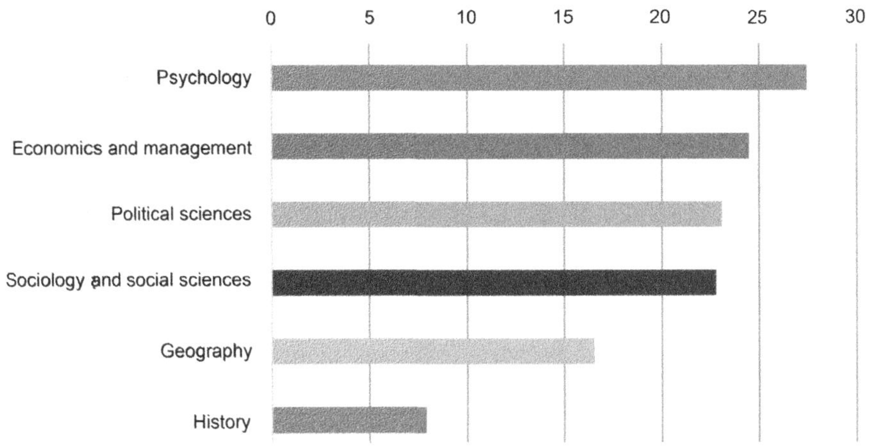


**Figure 5.8.** *Distribution of 6,970 occurrences of the term "resilience" by discipline (for the highest six rankings) as a percentage of the total found in the CAIRN database of publications in the humanities and social sciences (Robert 2019; source: www.cairn.info, accessed August 9, 2019). For a color version of this figure, see www.iste.co.uk/robert/resilience.zip*

### 5.2.3. *A controversial appropriation of the notion of resilience*

In 2007, researchers Fridolin Brand and Kurt Jax noted a tension between the scientific, descriptive concept of resilience as used in ecology and another, vaguer, more malleable meaning, which had emerged in the early 2000s, and which, in their opinion, posed a serious threat to the conceptual clarity and the practical relevance of the concept, given the high degree of normativity implied by some of its significations (Brand and Jax 2007). French geographers also expressed reticence with regard to the polysemic character of resilience, which emerged during the 2000s (Reghezza-Zitt *et al.* 2012; Reghezza-Zitt 2013). These limitations were a result of the positions taken at the moment the concept was adopted, but also of its evolutions within each

55 Seminar materials are available online on the *Résiliences urbaines – Département de géographie de l'ENS* website: http://geographie.ens.fr/-Resilience-urbaine-.html, accessed June 12, 2018.

discipline[56]. Resilience may appear as a new label for pre-existing concepts. The French psychologist Serge Tisseron, for example, considers that resilience was simply substituted for existing theories (attachment, vulnerability, etc.), and was less clearly defined than the concepts it replaced (Tisseron 2007, p. 120). In geography of risks, from the mid-2000s, resilience was increasingly present in discourse on the subject of disasters, alongside the more traditional concept of vulnerability. In this context, resilience was seen as a property which, implicitly, must be reinforced, as opposed to vulnerability, which has negative connotations and should be minimized (Lhomme 2012, p. 74).

The fact that resilience is seen as a positive attribute has not been properly discussed. In the context of urban development, Meerow *et al.* identified this assumption in 25 studies using the concept of resilience, which they then analyzed in detail (Meerow *et al.* 2016). As a systemic property, however, resilience should not form part of a value system. Furthermore, even systems that are considered to be undesirable may prove extremely durable (dictatorships, poverty, corruption, etc.) (Weichselgartner and Kelman 2015, p. 253). Resilience can maintain a system in a desirable or an undesirable state: this is the" dark side of resilience"[57]. While the maintenance of, or return to, an initial state may be perceived as resilience in the short-term, it may also hinder innovation and far-reaching transformations, which would increase long-term resilience (Quenault 2013a). Thus, the resilience of a system within a given duration or space may limit or run contrary to resilience on a different level. The geographer Patrick Pigeon has shown, for example, that the local settlement system of the town of Faute-sur-Mer, which appeared to be resilient at municipal level (low habitat density, dyke construction, etc.) was, in fact, vulnerable from a long-term perspective, as evidenced by the death of 29 inhabitants during Storm Xynthia in 2010 (Pigeon 2012).

Certain researchers have also noted that the focus on reactive capacities may limit attempts to cure the deep-seated vulnerabilities in a society. In the long-term, a refusal to take account of the permanent weaknesses of an urban system may lead to a reduction in their resilience (Gaillard 2010; Quenault 2013a). G. Djament-Tran notes that, in the second half of the 2000s, a form of "moralization of resilience" (*moralization de la résilience*) emerged, with a focus on "heroized" resilient individuals (*héroïsés*):

56 A useful comparison of different understandings of resilience across different disciplines can be found in Lhomme *et al.* (2010, p. 491).

57 Mitchell, T. and Harris, K., "Resilience: A risk management approach. ODI Background Note, January. London: Overseas Development Institute", 2012, https://www.odi.org/sites/odi.org.uk/files/odi-assets/publications-opinion-files/7552.pdf, consulted June 18, 2018, p. 5.

> Neo-liberal discourse concerning resilience forms a system with new scales of risk governance: the general tendency for the weakening of the state level in favor of the lower and higher levels can also be seen in the domain of resilience.[58] (Djament-Tran 2015, p. 64)

This type of call for resilience within populations may be seen as a means of shifting the responsibility for risk management onto populations themselves, and as a pragmatic response to state disengagement (Reghezza-Zit *et al.* 2012, p. 31; Quenault 2015). The connections between resilience and neoliberal politics are discussed at length in *The Routledge Handbook of International Resilience* (Chandler and Coaffee 2017).

The use of resilience is thus questionable in the context of human systems with their own sets of norms and values. The definition of "good" or "bad" resilience (Rufat 2012) or of desirable or undesirable states of a resilient system (when? For whom? etc.) raises questions of ideology and government approaches. The theoretical concept of resilience is watered down in an operational context, and risk management experts have tended to take a somewhat critical approach to its use.

### 5.2.4. *Assimilation with stability: engineering resilience*

Much of the criticism of resilience in operational contexts arises from an assimilation of this concept with that of stability (section 4.1). **Engineering resilience**, unlike ecological resilience (Holling 1996), focuses on short-term objectives that are easier to implement. It is characterized by constancy, predictability and control, with the overarching aim of maintaining the efficiency and yields of functions in a system, which is as stable as possible (Holling and Gunderson 2002, p. 27). Engineering resilience is dominant in the field of disaster management, particularly in France, where approaches tend to focus on reducing hazards (Quenault 2013a). M. Reghezza-Zit suggests that risk analysis and management has been dominated by a "technocentrist paradigm" (*paradigme technocentriste*) since the late 18th century. Thus, managers focus on the question of material damage and on the notion that technology can be used to manage catastrophes in a rational manner. Management approaches focus on increasing physical resistance and reducing exposure to risks by the establishment of protection

58 *Ces discours, fortement teintés de néolibéralisme, imposent de nouvelles échelles de gouvernance du risque : la tendance générale à l'affaiblissement de l'échelon étatique au profit des échelons inférieurs et supérieurs se manifeste ainsi aussi dans le domaine de la résilience.*

infrastructures, architectural norms and legal measures (regulation, zoning), etc. (Reghezza-Zit *et al.* 2012, pp. 38–39; Reghezza-Zit 2013, p. 3). Resilience, assimilated to stability, may be interpreted as resistance to change, as the goal is to ensure the maintenance of an initial "reference" state. This view clearly echoes the idea of an initial, final state, conserved over time, encountered in the domain of morphology (see sections 1.2.5 and 1.3.1).

The concept of resilience assimilated to stability, in the sense of engineering resilience, places considerable emphasis on shocks and ruptures, seen as exogenous perturbations that threaten the established equilibrium of a system. When resilience made its first appearances in the field of physics in France, in the 19th century, it was considered as one element of a dialectic alongside rupture. In 1901, G. Charpy invented a mechanism to measure resilience, which he called the *mouton-pendule* in reference to the *mouton* or *belier*, a medieval siege-engine comparable to the English battering-ram. The form of experiment used to measure resilience in physics involves the use of a violent rupture to measure a capacity of resistance. This reference to shock had implications on the way in which the concept was transposed into the field of psychology; psychologists acknowledged the physical origins of resilience (Cyrulnik 2002, p. 8; Vanistendael 2002). In psychology, just as in material resilience, the starting point for observation was marked by a shock or trauma (Cyrulnik 2002). Resilience was thus studied as a response to a violent accident in a post-catastrophic or post-traumatic context. In engineering resilience, the aim is to limit perturbations that may affect the stability of the system, while in the systemic view of resilience (ecological resilience), change and perturbations are an integral part of the system and contribute to its capacities to adapt (Holling 1986, p. 92) (section 4.2).

Engineering resilience has a number of limitations. First, it would be impossible to eliminate the risk inherent in the very nature of our societies (Beck 2001) and, secondly, it is hard to qualify changes as exclusively positive or negative, particularly with regard to their long-term effects. Generally speaking, in disciplines concerned with present-day human societies, the use of resilience raises ethical questions that are not found in metallurgy, ecosystems or in the study of past populations.

As early as 2001, members of the Resilience Alliance highlighted the need to specify the nature of resilience being measured: "resilience of what to what?" (Carpenter *et al.* 2001) and "for whom?" (Lebel *et al.* 2006; Cretney 2014). Researchers now recognize at least five different forms of resilience: "resilience for whom, what, when, where, and why". These distinctions are particularly applicable in the field of urban resilience (Meerow and Newell 2016). Geographers, too, have attempted to overcome the limitations of resilience in the context of human societies using the concept of specified resilience (Gunderson *et al.* 2010, pp. 34–35). In 2006, Walker and Salt defined two types of resilience: one specific and the other general. In the case of specific resilience, the part of the system and the type of perturbation in question must be defined. In general resilience, it is the system as a whole which is considered, with respect to perturbations of any type (Walker and Salt 2006). Specific resilience essentially applies to micro- and mesolevels and to the short-term, whereas general resilience focuses on the macrolevel and longer time periods (Toubin *et al.* 2012). An examination of the relations between different levels is particularly relevant in the context of the social sciences; as G. Djament-Tran has indicated,

> The critical reading of this rhetoric requires analyzing resilience not as a property of a level, but as a capacity to make the different scales interact with each other, to mobilize the lower and higher levels to facilitate the resilience process[59]. (Djament 2015, p. 64)

### 5.2.5. *Resilience as a systemic property: a limited impact*

Interactions between different levels are at the heart of the panarchy model put forward by C.S. Holling, L. Gunderson and G. Peterson in 2002, but the elements of the systemic theory of ecological resilience seem to have had a very limited impact on the work of French geographers, notwithstanding a detailed description in a 2007 article by A. Dauphiné and D. Provitolo. We carried out a search for the words *attracteur* (attractor), *bassin d'attraction* (basin of attraction), *cycle adaptatif* (adaptive cycle) and *panarchie* (panarchy) in 34 French theses and publications in which resilience was mentioned in the fields of geography and archeogeography between 2000 and 2017[60]. The notions of attractors and basins of attraction were the most widely cited, but even so, were only found in half of the studied publications. Furthermore, the term "attractor" may be used in a more common, general sense. The adaptive cycle and panarchy appeared in only one-third of the corpus, notably in explanatory contexts or critical analysis. There were few, if any, applications of these concepts to concrete geographical elements, which would enable these theoretical aspects of the conceptual framework of resilience to be truly tested (Table 5.2, Figure 5.9).

---

59 *La lecture critique de ces discours nécessite d'analyser la résilience non comme propriété d'un échelon, mais comme capacité à faire interagir les échelles, à mobiliser les niveaux inférieurs et supérieurs au profit du processus de résilience.*

60 Using articles and books in a digital format, our search is related to occurrences in the text, not including the bibliography or index, in singular and plural forms in both French and English.

| References | Attractor | Basin of attraction | Adaptive cycle | Panarchy |
|---|---|---|---|---|
| *Archaeomedes* 1998 | **3** | 0 | 0 | 0 |
| Aschan-Leygonie 2000 | 1 | **13** | 0 | 0 |
| Chouquer 2000 | 1 | 0 | 0 | 0 |
| Marchand 2000 | 3 | 0 | 0 | 0 |
| Chouquer 2003b | 1 | 0 | 0 | 0 |
| Robert 2003a | 4 | **15** | 0 | 0 |
| Beck *et al.* 2006 | 0 | 0 | **2** | 0 |
| Chouquer 2007 | 0 | 0 | 0 | 0 |
| Dauphiné and Provitolo 2007 | **15** | 8 | 0 | 0 |
| Maret and Cadoul 2008 | 0 | 0 | 0 | 0 |
| Garmy 2009 | 0 | 0 | 0 | 0 |
| Watteaux 2009 | 1 | 8 | 0 | 0 |
| Brigand 2010 | 0 | 0 | 0 | 0 |
| Lhomme *et al.* 2010 | 0 | 0 | 1 | **1** |
| Robert 2011 | 5 | 3 | 0 | 0 |
| Serre 2011 | 0 | 0 | 0 | 0 |
| Djament *et al.* 2012 | 0 | 0 | 0 | 0 |
| Lhomme 2012 | 0 | 2 | 0 | 1 |
| Pigeon 2012 | 0 | 0 | 0 | 0 |
| Robert 2012b | 2 | 0 | 0 | 0 |
| Toubin *et al.* 2012 | 0 | 0 | 0 | 0 |
| Barroca *et al.* 2013 | 0 | 0 | 0 | 0 |
| Quenault 2013a | 4 | 2 | 0 | 2 |
| Quenault 2013b | 0 | 3 | 7 | 5 |
| Reghezza-Zitt 2013 | 0 | 0 | 0 | 0 |
| Robert and Verdier 2014 | 3 | 0 | 0 | 0 |
| Garcia 2015 | 3 | 14 | **23** | **6** |
| Quenault 2015 | 0 | 0 | 0 | 0 |
| Reghezza-Zit 2015 | 0 | 0 | 0 | 1 |

| References | Attractor | Basin of attraction | Adaptive cycle | Panarchy |
|---|---|---|---|---|
| **Reghezza-Zit and Rufat 2015** | 1 | 6 | 2 | **6** |
| **Cavanna 2016** | 1 | 0 | 0 | 0 |
| **Chevillot-Miot 2017** | 1 | 0 | 6 | 4 |
| **Hermenault 2017** | 0 | 0 | 0 | 0 |
| **Robert 2017** | 0 | 0 | 0 | 1 |
| **Total** | 49 | **74** | 8 | 27 |
| **Frequency/34 references** | **13** | 9 | 7 | 9 |

**Table 5.2.** *Occurrences of terms belonging to the conceptual framework of ecological resilience (attractor, basin or domain of attraction, adaptive cycle and panarchy) in 34 publications by French geographers, archeologists and archeogeographers using the concept of resilience between 1998 and 2017. The earliest and strongest occurrences of each word are shown in bold (Robert 2019)*

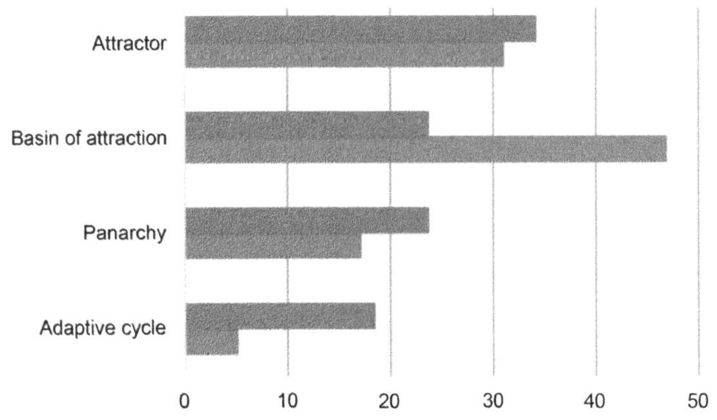


**Figure 5.9.** *Frequency and number of occurrences of the main concepts of ecological resilience in 34 publications by French geographers, archeologists and archeogeographers between 1998 and 2017 (Robert 2019, from data in Table 5.2). For a color version of this figure, see www.iste.co.uk/robert/resilience.zip*

COMMENT ON FIGURE 5.9.– *In terms of number (orange), the term basin of attraction is the most cited, but in terms of frequency (blue), attractor is the most widely used (34% of publications), outperforming basin or domain of attraction and panarchy (24%) and the adaptive cycle (18%).*

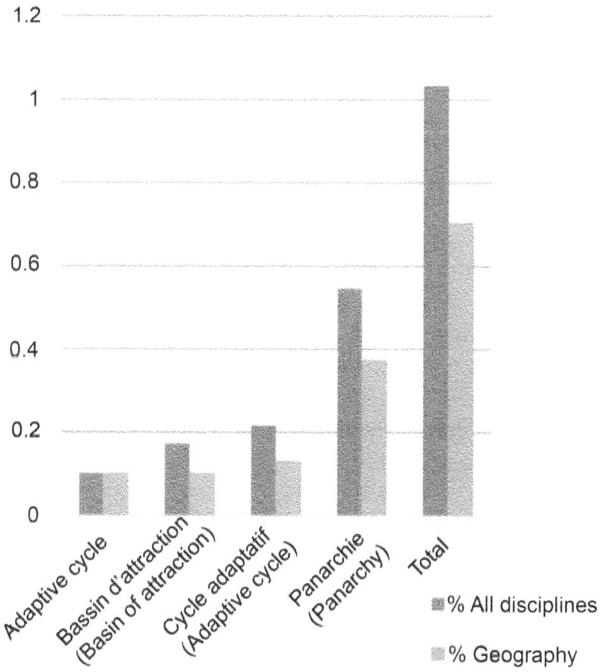


**Figure 5.10.** *Number of times the term "resilience" is cited alongside a concept associated with ecological resilience in the CAIRN platform, shown as a percentage of 6,970 occurrences. Note the very weak association of these theoretical concepts, accounting for less than 1.2% of the total. These terms are most widespread in geography (Robert 2019; source: www.cairn.info, accessed August 9, 2019). For a color version of this figure, see www.iste.co.uk/robert/resilience.zip*

A similar search of the CAIRN database also shows a weak adoption of the concepts of ecological resilience: out of 6,790 occurrences of the term "resilience", only 72 are associated with a theoretical element of the conceptual framework of ecological resilience (basin or domain of attraction, adaptive cycle or panarchy), representing 1.03% of the total[61] (Figure 5.10). However, interestingly, these occurrences display a different distribution across disciplines to that observed in Figure 5.8. In this case, the greatest number of citations of these aspects is found in geography (49 uses out of 72, i.e. 68%), more than in economics and sociology. Our search brought up no results for the field of psychology.

61 Research carried out using the AND Boolean: e.g. "resilience AND panarch*".

## 5.3. Conclusion

The concept of resilience was slow to be adopted in French geography, only truly making its appearance in the late 2000s and 2010s, after C. Aschan-Leygonie first highlighted the potential uses of the concept in 2000 (section 5.1.2.2). Essentially limited to the field of disaster management, it attracted a degree of criticism, notably regarding its use as a "discursive construct" (Reghezza-Zit and Ruffat 2015, p. xix), focusing on the way resilience has been used by public authorities and management bodies. Although M. Reghezza-Zitt and S. Rufat (Pierdet 2015) devoted a chapter to resilience as a systemic property in their 2015 work, it remains somewhat rare in the field. There are few empirical cases of application of the full conceptual framework of resilience (including the adaptive cycle, panarchy, alternative attractors, etc.). Nevertheless, of all the humanities, geography appears to be one of the domains in which the application of ecological resilience as a systemic property has progressed the furthest (Figure 5.10).

# 6

# The Conceptual Framework of Ecological Resilience: A Long-term Approach

## 6.1. The conceptual framework of ecological resilience: points for discussion

One of the difficulties of the conceptual framework of ecological resilience lies, in part, in the very definition of systemic resilience. A summary of the main definitions considered in this book is given in Table 6.1. C.S. Holling first defined resilience as a property of a system: the persistence of relations within a system and the capacity to absorb changes of state variables and parameters. This property enables a system to absorb and *utilize,* or even "benefit from" change (Holling 1973, p. 17, 1978, p. 11). Resilience may be measured in terms of the size of a basin of attraction, the strength of the repulsive forces at its boundaries, and the resistance to contraction of the basin itself (Holling 1986, p. 76). Resilience enables a system to absorb perturbations without tipping toward another attractor. This definition is applied in the analysis of alternative attractors[1]. Resilience corresponds to the size of the basin of attraction around a state, that is, the maximum number or size of perturbations that can be absorbed without a shift to an alternative attractor (Scheffer *et al.* 2001, p. 591, 2003, p. 650). For Walker *et al.* (2004) also, resilience is clearly defined as the maintenance of a system within its basin of attraction. However, in illustrating the stability landscape in which a system state shifts to an alternative basin of attraction following the contraction of the initial basin (Figure 4.3), the authors write: "Without itself changing, the system has changed basins" (Walker *et al.* 2004, p. 4). According to these definitions, resilience corresponds to the phases

1 In their analysis, Scheffer *et al.* explicitly state "Following Holling…" (Scheffer *et al.* 2001, p. 591) (section 4.1).

that precede a shift to another set of variables and interrelations, whereby a system enters a new region of stability (Holling 2006, p. 6).

Applying this vision to the adaptive cycle metaphor, resilience would correspond to the exploitation and conservation phases of the system: the former is characterized by high resilience, the latter by a progressive loss of resilience as the basin of attraction of the system begins to shrink, due to the increasing homogeneity of its constituent parts. However, in 1992, C.S. Holling also defined the resilience and recovery of a system in terms of sequences of release and reorganization (omega and alpha phases) in the adaptive cycle (Holling 1992, p. 450). These phases occur after the tipping point, which, in theory, determines the shift of a system into a new basin of attraction (and thus, potentially, an absence of resilience). In one case, resilience is seen as the capacity of a system to avoid shifting into a different regime; in the other, resilience is the capacity of the system to reorganize after a shift of this type.

Meerow *et al.* (2016) indicated that the way in which resilience is to be understood is dependent on the way system trajectories are defined. A system may exhibit persistence, resisting perturbations (engineering resilience), or it may adapt and undergo a progressive transition[2], or it may undergo a more radical transformation (Meerow *et al.* 2016, pp. 44–45). The lack of a clear definition of what, precisely, constitutes a transition in a system is a major source of ambiguity. Theoretically, a transition involves a shift to a different basin of attraction, and thus to an alternative attractor, resulting in the creation of a new system and the disappearance of the old system. However, certain systems may have two or more attractors, with the associated basins of attraction. How, then, do we define transition, and where should we look for resilience?

The role of endogenous and exogeneous (internal and external) factors is also a matter of debate. A system shift may be caused by one or more exogeneous perturbations, but, according to the adaptive cycle and alternative attractor approaches, the system itself also evolves, resulting in a reduction of its basin of attraction and increasing proximity to a tipping point. In this view, all systems are destined to disappear, which seems to imply that they cannot be resilient. C.S. Holling himself admitted that there was an unresolved question within the conceptual framework of resilience: what is it that enables the "rare transformations, not simply change" inherent in the life of a system (Holling 2006, p. 20)? On the subject of alternative attractors, Folke *et al.* wrote:

---

2. Other authors have used the term "transition" to describe a more radical transformation (section 4.5.1), whereas for Meerow *et al.* (2016), "transition" is less radical than "transformation".

> However, despite their elegance and rigor, they capture only part of reality. One of the main limitations of the dynamical systems theory that forms the broader underlying framework is that it does not easily account for the fact that the very nature of systems may change over time (Scheffer 2009). This implies that, in order to understand the dynamics of an intertwined social-ecological system (SES), other concepts are needed (Folke *et al.* 2010, np).

By taking account of different spatial and temporal scales, it thus becomes possible to specify the mechanisms of transmission within systems. In theory, the panarchy model offers a means of defining regime shifts and transitions in a more precise manner, introducing these different spatial scales and temporal rhythms. Resilience and transition can, then, coexist within the same entity; essential structural elements of a system may be maintained at one level, while major transformations are taking place on another level. Nevertheless, the possibilities of concrete applications of the panarchy model in the humanities and social sciences remain limited.

The adaptive cycle and panarchy models seem to involve a number of implicit assumptions, which have not been clearly demonstrated. Generally, the broadest spatial levels of social-ecological systems are associated with the slowest temporalities. The ecologist A. Kinzig notes that the idea that changes in slow variables erode resilience is based on intuition, rather than on demonstrated facts (Kinzig 2012, p. 317). The scalar models used in history and economics do, in fact, tend to assign the longest durations to large-scale organizations. Fernand Braudel, for example, referred to large-scale geographical frameworks, such as "the durable implantation of towns, the persistence of roads and traffic" as "prisons of long duration"[3] (Braudel 1958, pp. 731–732). However, cross-examining tables of physical and human geographical data with historical data from different temporalities, F. Durand-Dastès noted that the "diagonal effect" (*effet diagonal*) establishing a connection between large spaces and long temporalities and between small spaces and short temporalities is not always verified. Considering the global level, certain phenomena take place over very long periods (such as the evolution of *Homo Sapiens*, or long-term population movements); however, we also observe short-term phenomena with a significant impact (world wars, global crises, etc.) (Durand-Dastès 2004). The concordance between spatial and temporal scales still needs to be demonstrated in detail by means of multiple case studies.

3 *...la durable implantation des villes, la persistance des routes et des trafics.../prisons de longue durée.*

| Date | Definition | Reference |
|---|---|---|
| **Antiquity** | *Resilire:* "to leap or spring back; recoil; rebound; shrink (back again)" | Oxford 2005, p. 163 |
| **Medieval period** | "The action of retracting an act or one's word". | Godefroy 1881, p. 556 from an act of 1082 |
| **1901** | "The live resistance to rupture of an infinitely thin slice, or the work required to produce this rupture expressed in kilogrammeters per square centimeter of cross-section"[4]. | Charpy 1901, p. 873 |
| **1922** | "The effort required to break a bar". | Larousse 1922, p. 780. |
| **1973** | "There is another property, termed resilience, that is a measure of the persistence of systems and of their ability to absorb change and disturbance and still maintain the same relationships between populations or state variables". | Holling 1973, p. 14 |
| **1973** | "Resilience determines the persistence of relationships within a system and is a measure of the ability of these systems to absorb changes of state variables, driving variables and parameters and still persist. In this definition resilience is the property of the system and persistence or probability of extinction is the result". | Holling 1973, p. 17 |
| **1978** | "Resilience is a property that allows a system to absorb and *utilize* (or even benefit from) change". | Holling 1978, p. 11 |
| **1981** | "The measure of a system's, or part of a system's capacity to absorb and recover from the occurrence of a hazardous event". | Timmerman 1981, p. 21 |
| **1984** | "How fast the variables return towards their equilibrium following a perturbation". Calculated in units of time. | Pimm 1984, p. 322 |
| **1986** | "Resilience, on the other hand, is the ability of a system to maintain its structure and patterns of behavior in the face of disturbance. The size of the stability domain of residence, the strength of the repulsive forces at the boundary, and the resistance of the domain to contraction are all distinct measures of resilience". | Holling 1986, p. 76 |
| **1992** | "Those cycles are organized by four functions: exploitation, conservation, release, and reorganization. Stability and productivity are determined by the exploitation and conservation sequence. Resilience and recovery are determined by the release and reorganization sequence". | Holling 1992, p. 450 |

4 The original French text can be found in the Introduction to Part 2. Original versions of the citations in this table will only be given if they have not already been cited elsewhere.

| Date | Definition | Reference |
|---|---|---|
| **2001** | "The valley, or basin of attraction, around a state, which corresponds to the maximum perturbation that can be taken without causing a shift to an alternative stable state". | Scheffer *et al.* 2001, p. 591 |
| **2003** | "The size of the basin of attraction is also called resilience". | Scheffer *et al.* 2003, p. 650 |
| **2004** | "Resilience is the capacity of a system to absorb disturbance and reorganize while undergoing change so as to still retain essentially the same function, structure, identity, and feedbacks — in other words, stay in the same basin of attraction". | Walker *et al.* 2004, p. 6 |
| **2006** | "Hence the useful measure of resilience was the size of stability domains, or, more meaningfully, the amount of disturbance a system can take before its controls shift to another set of variables and relationships that dominate another stability region". | Holling 2006, p. 6 |
| **2017** | "Resilience is the capacity of a system to absorb disturbance and reorganize while undergoing change so as to still retain essentially the same function, structure, identity, and feedbacks — in other words, stay in the same basin of attraction". | Resilience Alliance 2017 |

**Table 6.1.** *Summary of the different definitions of resilience explored here (Robert 2019)*

The same may be said of the relations established between different levels in the panarchy model. In the representation proposed by Holling *et al.* (2002), the interscalar arrows appear to be systematically associated with specific levels: the large, slow level in the case of the "remember" arrows, and the small, fast level for the "revolt" arrows. There are no "revolt" arrows associated with the large, slow cycle, nor are there any "remember" arrows for the small, fast cycles. Nevertheless, this specialization of levels is never really discussed, and the reverse may, in fact, be true in certain cases (cf. our discussion of cultural landscapes in section 6.3).

The failure to take account of time may be seen as a major limitation of the proposed models. A long-term approach to system observation offers a means of defining thresholds or transitions more clearly, and of specifying rhythms and temporalities as a function of levels.

## 6.2. The benefits of the long-term perspective

As early as 2001, archeologist Sander Van der Leeuw and geographer Christina Aschan-Leygonie highlighted the interest of a long-term approach to the resilience of social-ecological systems in their article *A Long-Term Perspective on Resilience in Socio-natural Systems*, published by the Santa Fe Institute (Van der Leeuw and Aschan-Leygonie 2001). Drawing on work carried out as part of the *Archaeomedes* program, they demonstrated the interest of historical and archeological cases for relativizing the notion of an "environmental crisis".

In her study of the Comtat plain, C. Aschan-Leygonie compared two agrarian crises which occurred approximately one century apart (Aschan-Leygonie 1998). Between 1860 and 1890, farmers in the Comtat were forced to abandon their traditional crops due to competition (in the case of madder and silk) or to destruction (grape vines), and developed arboriculture and horticulture based on an inherited network of irrigation canals, small farms and fragmented plots. This legacy meant that the Comtat farming community was able to establish a new system without needing to mobilize significant capital or human resources. The interaction between different levels played a decisive role: at a regional level, the construction of the Paris-Lyon-Marseille railroad provided the infrastructure required for this new form of agriculture and for the transportation of seasonal laborers. For Aschan-Leygonie, this example demonstrates how, "following a perturbation, the system is not marked by a return to equilibrium, expressing resistance, but, on the contrary, reacts in a manner which is often positive and creative manner, through a series of changes and adjustments"[5] (Aschan-Leygonie 2000, p. 65).

However, in the 1970s, production in the Comtat ceased to be competitive, and the crisis deepened as markets were opened in the context of the expansion of the European Union. The small size of individual farms and plots, low levels of investment, etc., limited the acquisition of new equipment. The existing irrigation system was no longer sufficient, and any attempts to transform plot layouts were hindered by the presence of inherited paths and hedges. Aschan-Leygonie's study clearly highlights the difference between stability and resilience. In the first case, the system was destabilized (rapid disappearance of traditional crops), but demonstrated resilience. In the second case, the system was relatively stable, but not particularly resilient. Inherited structures play different roles in different contexts: "Inherited structures either hinder the evolution of a system by their rigidity, or contribute to its

5 *...après une perturbation, le système n'est pas marqué par un retour à l'équilibre, expression d'un comportement de résistance, mais réagit au contraire de manière souvent positive, créatrice, grâce à de multiples changements et réajustements.*

dynamics"[6] (Aschan-Leygonie 2000, p. 75). In the 19th-century case, inherited structures contributed to the resilience of the system, while in the 20th century, they appear to have hindered the adaptation of local societies to a crisis[7].

Given the impossibility of predicting and controlling all perturbations, Aschan-Leygonie emphasized the notion of learning, notably **learning from the past**. She stated:

> Among other things, it is by learning from the effects of past perturbations that the system will be able to respond to change in an active manner. This response capacity is linked to both individual and institutional perception, and to individual and collective memory[8]. (Aschan-Leygonie 2000, p. 71)

Awareness of a system's past, notably experiences concerning the effects of prior perturbations, thus enables actors to accumulate knowledge with respect to change, and to respond in a faster, more active manner to future changes.

In 2003, the anthropologist Charles Redman and the ecologist Ann Kinzig published an article on the interest of a *longue durée* approach[9] to documenting interactions between the different scales involved in systemic resilience (Redman and Kinzig 2003). They called for an increased participation of the archeology community in debates on resilience[10], as at least four fundamental elements are found in the discipline.

1) While researchers working on present-day situations generally see only a "truncated" form of the adaptive cycle, archeology can "fill some of the gap", in some cases providing a vision of multiple, complete cycles. It also provides the tools for observing slow processes, which are essential to the resilience of a system and which cannot be understood by studying the present alone.

2) Archeology offers a means of observing the long-term relevance of response and strategies developed by societies in response to past environmental crises.

---

6 *Soit les structures héritées entravent l'évolution du système par leur rigidité, soit elles servent sa dynamique.*

7 One could object that, in the first case, the consequences of perturbations can be seen over more than a century, while for the second case (beginning in the 1960s), the period examined was much shorter – around 30 years.

8 *...entre autres, par l'apprentissage des effets des perturbations passées que le système parviendra à répondre activement aux changements. La capacité de réponse sera liée à la fois à la perception des individus et des institutions et à la mémoire individuelle et collective.*

9 In French in the original text.

10 Cf. Van der Leeuw and Redman (2002).

Archeologists may show that certain responses which appeared effective in the short-term led to a serious reduction in long-term resilience, for example in terms of soil degradation. Certain solutions, which may appear illogical from a short-term perspective, may thus be shown to reduce risk and increase long-term resilience.

3) Archeology makes it possible to identify "true transformations", such as the development of agriculture, urban settlements and industrial societies, based on multiple observations across different temporal and spatial levels.

4) Archeology and anthropology provide the tools for studying the interrelations between human social systems and ecosystems from the perspective of different organizational levels (from habitat to the society as a whole), improving our understanding of the interactions between slow and fast variables.

The long-term perspective is thus key for documenting interactions between the different levels at work in panarchic system dynamics. Redman and Kinsig's article inspired a series of other works, confirming the role of archeologists in the management of contemporary socio-natural systems, using the theoretical framework of resilience.

In 2005, the *American Anthropologist* journal published a dossier entitled *Landscapes over Time: Resilience, Degradation, and Contemporary Lessons* (Fisher and Feinman 2005). The authors considered landscape as a laboratory for observing the long-term relations between humans and their environment, working on different periods and cultural spheres; they highlighted the interest of the historical trajectories of different sites in understanding present degradations (desertification, deforestation, soil degradation, etc.). The editors of the dossier, Christopher T. Fisher and Gary M. Feinman, called for the unification of studies carried out across different disciplines, on different themes, different periods, etc., bridging the gap between contemporary and historical studies. In this way, archeology and the historical sciences as a whole could play a role in current environmental debates. They highlighted the need for a common, transdisciplinary language (metalanguage), drawing on key concepts from the fields of historical ecology, political ecology and agro-ecology. These concepts included resilience and that of a "legacy of human capital" (Fisher and Feinman 2005, p. 64).

The adoption of resilience in archeology occurred simultaneously to that of another concept, borrowed from the field of historical ecology: "landesque (or laboresque) capital", which is essentially the capital transmitted by working the land. Beginning in the 1980s and 1990s, researchers worked on demonstrating the potential of old techniques or types of agrarian organization for present-day economic development (Altieri 1995; Denevan 1995). They showed that long-term development of a landscape by the societies inhabiting it can generate gains in productivity in the present. Irrigation or drainage networks, terraces and areas of cultivation, developed and maintained over long periods of time, form a type of built capital that may contribute to the productivity of present and future generations (as in the case of the Comtat: section 6.2). The concept of *laboresque* or *landesque capital*[11] was first introduced by the economist Amartya Sen in the context of agricultural development in developing countries; in the 1980s, it was adopted by the English geographer Harold Brookfield in his work on agrarian organizations in Oceania (Brookfield 1984). In these approaches, landscape and resources are seen as the product of interactions between societies and the environment over long periods. They are constructed over time, and, to some extent, decisions made in the past condition present and future landscapes.

In 2009, Fisher *et al.* (2009) published the results of work carried out over a period of several years in *Exploring Solutions Past: The Role of Archaeology in Shaping Conservation Strategies of the Future. The Archaeology of Environmental Change: Socionatural Legacies of Degradation and Resilience. Various contributions, such as S. Van der Leeuw's What is an "Environmental Crisis"?*, showed how archeologists can actively contribute to debates over the adaptation of societies to environmental change, both by providing a finer reconstruction of system trajectories and by observing how past societies dealt with climate change and environmental crises (Van der Leeuw 2009). By the late 2000s, collaborative studies by researchers working on both past and present were becoming increasingly common, focusing on the behavior of ecosystems under stress; for example Braje *et al.* (2009) and Campbell and Butler (2010) worked on the exploitation of fishing resources.

In 2011, an issue of *Ecology and Society* focused on the contributions made by archeological studies in testing propositions intended to improve the resilience of current-day social-ecological systems. The authors presented a retrospective study of strategies developed by human societies in order to adapt to change (e.g. resettlement, diversification of resources, increased vulnerability on certain levels, and so on) and considered ways in which past societies might have avoided or attenuated the effects of "traps" that forced them into significant and undesirable transitions (Schoon and Fabricius 2011).

Finally, in *Social Sustainability, Past and Future. Undoing Unintended Consequences for the Earth's Survival*, S. Van der Leeuw presented a summary of several years' work concerning the contributions of the long-term perspective in analyzing the resilience of social-ecological systems, highlighting the close connections between past, present and future. The long-term perspective notably gives us an improved understanding of slow dynamics of both natural and social origin that affect present-day systems. Historical and archeological studies can also

11 Amartya Sen, cited in Håkansson and Widgren (2014, p. 9).

be used to define pre-perturbation states of reference, sometimes revealing reasons why a system might not be viable over time. Finally, they cast light on the transformations affecting social-economic systems, but also on the processes of change themselves (the change of the change). Van der Leeuw suggested an interpretation of the adaptive cycle in terms of energy and information flows, two elements considered as drivers of long-term societal resilience (Van der Leeuw 2020).

## 6.3. The cultural landscape: a field of heuristic experimentation

### 6.3.1. *The contribution of cultural landscapes and land-use legacies*

From the 2000s, the question of landscape was perceived as a field of heuristic experimentation for testing the conceptual framework of resilience, through the observation of social-ecological systems from a long-term perspective. Landscapes are one of the keys to understanding the relation between humanity and natural resources, notably through the notion of **land use**. This idea was particularly prevalent in the field of historical ecology in the 1990s–2000s (Szabó 2014). The archeologist Carole Crumley highlighted the role of the French *Annales* school, among other intellectual currents, in the emergence of this idea, notably in terms of attempts to formalize long-term observations (Crumley 2012). As a field of research, historical ecology has a strongly interdisciplinary character[12], exploring concrete recommendations for land and landscape management.

The early 2000s saw a *rapprochement* between landscape archeology and landscape ecology, focused on the idea that land use has a long-term impact and leaves an ecological legacy (ecological land use legacies) (Briggs *et al.* 2006; Rhemtulla and Mladenoff 2007). The way in which resources have been used may have a lasting effect on ecosystems, influencing biodiversity and current productions, and may even limit possible future responses. For example, studies have aimed to demonstrate the long-term impact of resource usage on ecosystemic services. While some usages promote open, diverse trajectories, others can lead to blockages (legacy lock-ins) (Ziter *et al.* 2017). These legacies have been studied in the context of forest, aquatic and arable cultivation, among others[13]. In these approaches, past and present forms are analyzed together in order to reconstruct system trajectories. Both archeologists and ecologists use tools such as geophysics, teledetection, LiDAR (Light Detection and Ranging), etc. In Caracol (Belize), for example, the biologist John Weishampel studied the effect of terraces constructed by Mayan populations on the

12 See the example of G. Bertrand, described above, combining geography, ecology, archeology and history (section 3.1.2).

13 See Foster *et al.* (2003), Perring *et al.* (2015) and Bürgi *et al.* (2017).

present-day structure of the forest. Although these terraces have not been maintained over the last thousand years, their presence has an impact on the present form of the forest canopy (Weishampel *et al.* 2013).

The field of historical ecology, bringing together ecologists, archeologists, historians, anthropologists and geographers, is a fertile ground for the use and critical analysis of the conceptual framework of resilience. Cultural landscapes provide a concrete focus for experimentation. The notion of **cultural landscape**, used by 19th-century geographers then adopted by Carl Sauer in the early 20th century, or in the sense used by UNESCO[14], presumes the existence of interactions between nature and society. For Thomas Kirchhoff *et al.*, the notion of a social-ecological system has strong similarities to the concept of cultural landscape, developed from the 1970s on and based on the same conception of the relations between humanity and its environment (Kirchhoff *et al.* 2012).

In 2012, a group of researchers brought together by German ecologists Tobias Plieninger and Claudia Bieling experimented with the connections between the conceptual framework of cultural landscapes and that of resilience, as developed by the Resilience Alliance. The dynamics of past and present landscapes were analyzed, forming the basis for a discussion of potential orientations for the sustainable management of social-ecological systems (Plieninger and Bieling 2012). For A. Kinzig, one of the book's contributors, the interest of cultural landscapes lies in highlighting the social approach to research on social-ecological systems (Kinzig 2012). Throughout the book, real-world cases are presented as means of testing the adaptive cycle and panarchy models. The authors showed that the succession of the four phases in the adaptive cycle was not, in fact, inevitable; several phases can exist simultaneously, and the phase of collapse may be avoided. Studying the resilience of the landscape created by the sugar cane industry in the Caribbean, for example, William Found and Marta Berbès-Blàzquez identified at least four elements that indicate weaknesses in the conceptual framework of resilience (Found and Berbès-Blàzquez 2012):

1) Several different regimes may coexist on an island or group of islands, and do not fluctuate in response to changes in external conditions, in contrast to the transitions observed in natural systems. Regime change occurs gradually.

2) Several adaptive cycles may be at work on the same level, but in the panarchy model, only one cycle is represented at each level. The authors proposed alternative representations of panarchy featuring up to four levels (Figure 6.1).

14 Cultural landscapes are defined here as "combined works of nature and humankind". "Cultural Landscapes". UNESCO World Heritage Centre. https://whc.unesco.org/en/culturallandscape/, consulted July 3, 2019.

3) "Revolt" and "remember" arrows do not originate in a single level, as in the initial model proposed by Holling *et al.* (2002), but may emanate from slow or fast levels. In the cited example, variables decided by global institutions may lead to rapid transformations in the system on the smallest scale, that of individual property. Inversely, phenomena of memory or resistance may have their origins in the broadest, slowest levels (global institutions), but also in intermediate levels (island level). The authors cited cases where major landowners resisted innovations brought in by the local authorities. In other cases, agricultural innovations originating at plantation level were limited by climatic constraints (global level)[15]. The authors proposed a modified version of the panarchic model for their case study (Figure 6.1).

4) The adaptations of the "sugar cane landscape" in the Caribbean do not appear to have included rapid or catastrophic phases; change seems to have been incremental, involving actors at different levels. Nevertheless, the 19th-century system of large, specialized plantations was very different to the pattern of exploitation observed for the late 16th century. The landscape resulting from the sugar cane industry can be seen to be more resilient than the production system which created it (Found and Berbès-Blàzquez 2012, p. 176).

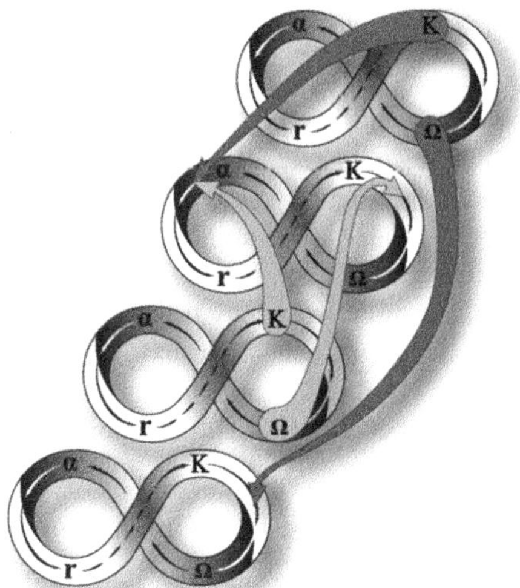


**Figure 6.1.** *Interscalar interactions in the panarchy model for sugar cane production in the Caribbean (adapted from Found and Berbès-Blàzquez (2012, p. 180) by S. Robert, 2019)*

15 "Innovative and stabilizing forces can originate from fast- or slow-moving variables, the process of propagation or resistance can be bottom-up or top-down, and interactions between elements in the panarchy can skip levels" (Found and Berbès-Blàzquez 2012, p. 168).

COMMENT ON FIGURE 6.1.– *Remember arrows (dark gray) and revolt arrows (light gray) emanate from different levels: this was not the case in the original model proposed by Holling* et al. *(2002). These arrows represent innovations and resistances at different levels, such as landowner resistance to transformations imposed from above, or climatic constraints (macrolevel) limiting transformations instigated at plantation level.*

The proposed models of cultural landscapes tend to include greater numbers of potential interactions between different levels, with a broader heterarchy (diversification of possible relations) characterizing change in systems (Crumley 2012). The articulation between the conceptual frameworks of resilience and political ecology was an area of debate, notably with regard to the interactions between different levels (Widgren 2012; Turner 2014). These approaches were tested in anthropology, in the context of applied programs such as the REAL (Resilience in East African Landscape) project, coordinated by Paul Lane[16] and financed as part of the European Commission's Seventh Framework Program between 2013 and 2017. The program involved close collaboration between ecologists, archaeologists, anthropologists, geographers, historians and agronomists, with the aim of training a new generation of ecosystem service managers, political decision-makers and entrepreneurs[17].

### 6.3.2. *The role of spatial markers in social resilience*

A further promising field for landscape studies concerns the active role which it may play in the resilience of settlement systems, via its capacity to transmit memory through spatial markers. Archeologists Victor D. Thompson and John A. Turck, studying the reoccupation of coastal land in the present-day state of Georgia (USA) by hunter-gatherers, in the wake of major sea-level fluctuations between 4200 and 1000 BP, highlighted the role which may be played by traces of former occupation. This study was particularly interesting in its focus on the relations between a population, a particular resource type and a territory through a full adaptive cycle (Thompson and Turck 2009).

The Late Archaic period (4200–3100 BP) appears to have seen a first intensive occupation of the coastal area by hunter-gatherer groups, dependent on a coastal and estuary-based economy focused on fishing and the harvesting of shellfish. Dwelling sites are notably characterized by collections of shells in the form of a circle or U

16 Professor of global archaeology at the University of Uppsala.

17 http://www.real-project.eu/, accessed January 24, 2018. For the French context, see the work of Benoît Hazard (*Institut interdisciplinaire d'anthropologie du contemporain* – EHESS – CNRS): http://www.iiac.cnrs.fr/article453.html.

around a central area, known as shell rings. According to Thompson and Turck's interpretation, the Late Archaic period, corresponding to the implantation and maintenance of a population along this coastal area, may be assimilated to the r and K phases of the adaptive cycle (development and conservation). The K phase is characterized by a strong connection between the society and fishing resources; this creates a certain vulnerability to variations in these resources. During the subsequent Early Woodland period (3100–2400 BP), a significant coastal regression affected the whole ecosystem of the estuary marshlands. Variations in soil salinity affected the distribution of the shellfish which constituted the main component of the hunter-gatherers' diet. This period saw a significant decline in population, accompanied by transformations in terms of distribution. There are no major shell deposits from this time, and the islands in the delta barrier were abandoned. Thompson and Turck suggested that this episode, traditionally viewed as a collapse in the sociopolitical integration of the Late Archaic period, corresponded to the release (Ω) phase of the adaptive cycle. Groups adapted to environmental changes by adopting a more mobile lifestyle and by diversifying their resources, notably increasing their use of terrestrial, rather than maritime, resources through the development of hunting.

During the Middle Woodland (2400–1500 BP) and Late Woodland (1500–1000 BP) periods, sea levels returned to their previous state, and the density and extension of settlement returned to levels comparable to those of the Late Archaic period. Intensive use of coastal resources resumed, although populations continued to hunt. Previously abandoned lands were reoccupied. The shell rings left by Late Archaic groups appear to have acted as focal points for several new settlements. Although the scale of shell deposits from this period never matched that of the Late Archaic period (perhaps because the resource itself failed to reach previous levels), groups of settlers appear to have behaved in a similar, albeit increasingly complex, manner, as indicated by the appearance of tumuli (funerary structures). These have been interpreted as a way for certain family groups to assert a form of control over resources and their desire to protect these resources for future use. Architectural traces may contribute to societal resilience by transmitting a form of social memory, notably memory of possible vulnerabilities with regard to perturbations affecting coastal resources. This memory transforms the perception that new settlers may have of their environment (Thompson and Turck 2009). Thompson and Turck did not illustrate the adaptive cycle they described in their text; the diagram shown below (Figure 6.2) was produced by drawing on Thompson and Turck's work and the theoretical representation developed by Holling (1986) and Gunderson and Holling (2002).

Cultural practices, particularly architectural traces which bear witness to the relations between men and their environment, contribute to the resilience of societies by transmitting a form of memory. This memory does not prevent fluctuations, but prevents the emergence of an entirely new system. The French geographer Antoine Le Blanc notes:

> Resilience has thus been strongly linked to regional identity, to local knowledge that is habitual, lived, felt and known; resilience increases proportionally to better knowledge of the region and its inhabitants, to the quality of the relationships and interactions between stakeholders. [...] Shared memory constitutes the basis of a district's identity, and resilience is based on territory and memory. The memory of place, the functioning of a regional system and regional habits, contributes to consolidating places' and inhabitants' resilience.[18] (Le Blanc 2015, pp. 189–190)

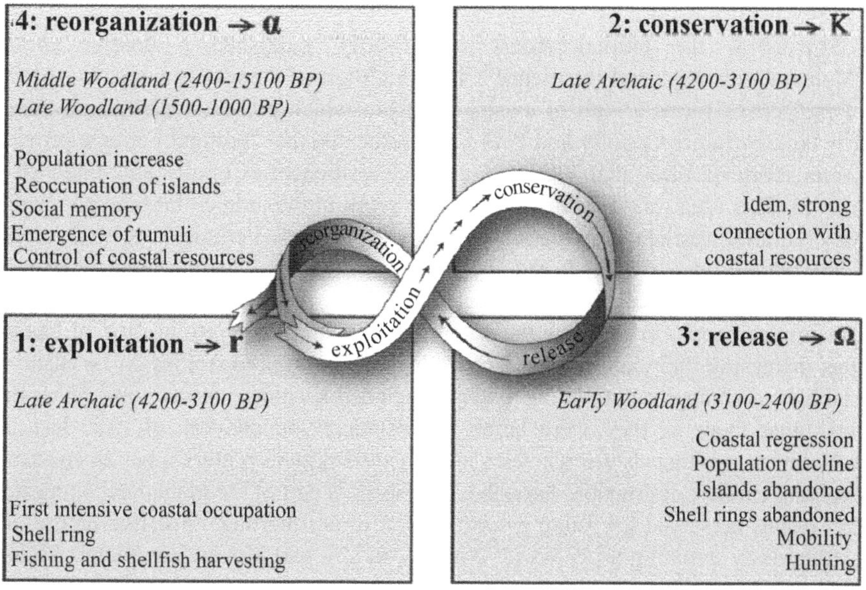


**Figure 6.2.** *Proposed reconstruction of the adaptive cycle described by V. Thompson and J. Turck for hunter-gatherers in Georgia (USA), based on the adaptive cycle model proposed by (Holling 1986; Gunderson and Holling 2002) (S. Robert, 2019)*

18 *La résilience est ainsi fortement liée à l'ancrage territorial, à la connaissance du lieu, habité, vécu, ressenti, connu ; la résilience augmente de manière proportionnelle à la meilleure connaissance du territoire et de ses habitants, à la qualité des relations et interactions entre acteurs. [...] La mémoire commune constitue le socle de l'identité d'un territoire, et la résilience repose sur cet ancrage territorial et mémoriel. La mémoire du lieu, du fonctionnement du système territorial, les habitudes contribuent à consolider la résilience des lieux et des habitants.*

The temporality of crises is, in fact, relatively short, and is dependent on individual perceptions (Van der Leeuw 2009). The projection of the memory of past crises onto monuments, high-water marks, etc., preserves the memory of a catastrophe outside of the population of individuals who experienced it. A. Le Blanc referred to the conservation of ruins associated with trauma as "a kind of urban post-it note, an instrument for memory and resilience"[19], notably in the case of Italian towns which had experienced major earthquakes (Le Blanc 2015, p. 191). This memory is seen as a tool for "proactive" resilience, contributing to a reinforcement of the notion that learning from the past can increase the resilience of societies (Aschan-Leygonie 2000) (section 6.2).

In 2018, the journal *World Archaeology* published a dossier entitled "Vulnerability, Risk and Resilience". They highlighted the way in which knowledge of past crises forms a well of experience from which present societies may draw. The editors, James Conolly and Paul Lane, indicated that "cultural heritage provides a repository of ideas that increases cultural resilience" (Conolly and Lane 2018, p. 548). This idea was developed in greater depth in the article "Embracing change: how cultural resilience is increased through cultural heritage" by archeologist Cornelius Holtorf. He expressed the notion that cultural heritage may increase a society's capacity to embrace change: "The evident changes of heritage over time can inspire people to embrace uncertainty and absorb adversity in time of change, thus increasing their cultural resilience" (Holtorf 2018, p. 639). Like A. Le Blanc, he highlighted an approach to heritage which included keeping ruins as ruins, rather than rebuilding them as they were before, to preserve the memory of risk. Just like landscapes, monuments are not seen as a finished state to preserve, but as constantly evolving entities; destruction, through catastrophe, is part of the monument's trajectory and is thus preserved for future generations: "Cultural heritage, just like nature, is a continuously evolving process, not a legacy in any part already complete" (Holtorf 2018, p. 644).

## 6.4. Transitions in settlement systems

Another field with the potential to enrich the conceptual framework of resilience is the study of settlement systems in the long term. Focusing on the structures which people have created for shelter, circulation and the obtention, exploitation and management of resources, the study of settlement systems offers a vision of variations of population distributions in a given territory[20]. This field of study has

19 ...*manière de* post-it *urbain, instrument de mémoire et de résilience.*

20 Definition based on descriptions found in Pinchemel and Pinchemel (1988), Brunet *et al.* (1995) and Sanders (2017, p. 34).

formed the basis for multidisciplinary programs involving archeologists, geographers, ecologists and more, notably in relation to understanding transitions (section 5.1.2.2).

Beginning in the second half of the 2000s, researchers began to apply Holling's adaptive cycle model to characterize transitions observed in indigenous North American villages between the classical and post-classical periods. They showed that phases of depopulation and habitat dispersal, traditionally considered to represent the destructuring of indigenous societies, may also be interpreted as release (Ω) phases, followed by a phase of reorganization (Nelson *et al.* 2006; Redman *et al.* 2009; Thomson and Turck 2009; Nelson *et al.* 2011). Studying pre-Hispanic populations in the southwestern United States, a group of archeologists and ecologists led by Michelle Hegmon *et al.* attempted to identify rigidity traps, theoretically characterized by increased connectivity alongside a decline in diversity and innovation, factors likely to lead to a decline in resilience and to transition toward a new attractor (section 4.3). Signs of this "rigidification" of the system were observed in a variety of variables (population displacement, transformations in terms of technology, architecture, burial practices, traces of violence and physical suffering, etc.). The authors argued that the hierarchization of society, increased interdependence of activities and a loss of diversity, in both cultural practices and innovation, may be seen as a rigidification of the society in question, with a subsequent loss of resilience leading to a potential transition (Hegmon *et al.* 2008).

It is thus possible to understand questions of societal collapse within the context of settlement systems. The term "collapse" is traditionally used in this context to describe:

> the fragmentation of states into smaller political entities; the partial abandonment or complete desertion of urban centers, along with the loss or depletion of their centralizing functions; the breakdown of regional economic systems; and the failure of civilizational ideologies. (Schwartz and Nichols 2006, pp. 5–6).

The term is generally used to define the end of complex, hierarchized and organized societies (Diamond 2005; Schwartz and Nichols 2006; Rice 2013; Faulseit 2016). The concept of resilience shifts the focus onto the means which populations develop – notably in terms of mobility – to adapt when faced with limited access to resources, and with changes which may be rapid and unpredictable. Prudence Rice (2013) explains that this is essentially a question of perspective: "glass half-full versus glass half-empty" (p. 17). The concept of resilience is used alongside others, such as that of regeneration, which defines the reappearance of complex societies following periods of decentralization, or that of survival, via the maintenance of certain elements of symbolic importance, such as language, by elites (Schwartz and Nichols 2006, pp. 8–10; Weiberg 2012; Rice 2013). The book *Beyond*

*Collapse: archaeological perspectives on resilience, revitalization, and transformation in complex societies* provides a detailed examination of alternative concepts, such as resilience or revitalization, which may be substituted for "collapse" (Faulseit 2016).

In *Social Sustainability, Past and Future*, S. Van der Leeuw suggested that major transformations observed in different civilizations might be seen as different stages in the system of information processing in societies, moving from egalitarian organizations to hierarchical urban structures[21]. He described what he saw as the four major bifurcations in the history of a society: the emergence of information-processing hierarchies in egalitarian societies; the emergence of proto-urban geographic centers from which information is disseminated (such as early medieval emporia in Northern Europe); the interconnection of centers along relatively stable long-distance transport corridors (as seen in La Tène culture in Europe, or in certain phases of Mayan, Chinese and Near-Eastern civilizations); and, finally, the growth of towns that are highly dependent on their supply and trading networks (Van der Leeuw 2020, pp. 263–287).

Work on urban geography in France has also contributed to our understanding of transitions. Researchers have drawn extensively on the formalization of the dynamics of spatial systems proposed by F. Durand-Dastès (section 5.1.2.1). Comparative statistical models have been developed to compare and contrast successive descriptions of spatial situations:

> Fundamentally, these consist of producing representations of the successive states in an evolution, aiming to establish the rules of operation of each state – the synchronous perspective, or the level 1 domain of temporality – and of indicating indicate the ways in which this state is at least partly conditioned by a previous state – introducing diachrony, by means of level 2 temporality[22] (Durand-Dastès 1999, p. 14)[23].

---

21 Hybrid town systems combining both egalitarian communities and more complex hierarchies (Van der Leeuw 2020, p. 200).

22 *Fondamentalement, ils consistent à produire des représentations des états successifs d'une évolution, étant entendu qu'à chaque étape on cherche à poser les règles de fonctionnement de l'état, – point de vue synchronique, ou domaine de la temporalité de niveau 1 – et à indiquer comment cet état est conditionné au moins en partie par un état antérieur – introduction de la diachronie par la temporalité de niveau 2.*

23 Such as the "principles" described by Christian Grataloup (1996) or Brunet's (2001) "choremes".

Other types of formalization can be used to describe the aspects of a system: variations in a quantity (such as the population of a district or town) as a function of several internal or external variables, using differential equations (Sanders 1992), cellular automata or multi-agent systems, for example. The latter case focuses on the way in which the behaviors of "'agents', with capacities of autonomy, adaptation and interaction"[24] (Sanders 2007) combine to produce perennial, and thus observable, macrostructures. Durand-Dastès states: "From a spatial perspective, we go from small to large; in temporal terms, we work from the short temporalities of interindividual relations up to more or less stable structures, considering the conditions of their emergence"[25] (Durand-Dastès 2004, p. 21).

Following on from the work of the *Archaeomedes* program, the *TransMonDyn* program, organized by the *Agence nationale de recherche française* and directed by Lena Sanders, brought together geographers, archeologists, historians, linguists, mathematicians and computer scientists, working on questions of transitions affecting settlement systems (Sanders 2017). The authors drew on formalizations of transitions and regime shifts used in mathematics, physic, ecology and computer science to develop a common conceptual framework for the description of transitions observed in 12 different case studies. The subjects of these studies ranged from global to micro-regional scale and from prehistory to the present day, covering durations from a few decades to millennia (Sanders 2017, pp. 33–53). Significant work was carried out on specifying ontologies in order to create a shared descriptive framework, making it easier to model the dynamics at work in different systems.

Terms such as "transition", "bifurcation" and "resilience" formed part of the lexicon used in geography and archeology, defined in greater detail by the *Systèmes de peuplement sur le temps long* (Long-term settlement systems) working group at the *Laboratoire d'excellence "Dynamiques territoriales et spatiales"* (Center for Excellence, "Territorial and Spatial Dynamics": LabEx DynamiTe), hosted by the *Agence nationale de la recherche française* (Sanders *et al.* 2020b). Two specific chapters were devoted to this subject (Sanders 2020; Robert 2020).

## 6.5. Conclusion

As a discursive construct, resilience suffers from a number of imprecisions and limitations. As a systemic property, however, resilience has acquired a precise theoretical foundation in the field of ecology, and has been enriched by

24 … *"d'agents" dotés de capacités d'autonomie, d'adaptation et d'interaction…*

25 *Du point de vue spatial, on chemine du petit vers le grand, du point de vue des temporalités, on part des temps courts des relations interindividuelles pour aboutir aux structures plus ou moins stables, et on s'intéresse aux conditions de leur naissance.*

contributions from other scientific disciplines. This systemic concept goes beyond the traditional nature/culture dichotomy; society/environment relations are presented as a complex social-ecological system. The conceptual framework of ecological resilience bridges the gap between the life sciences and social sciences, and between the sciences of the past and present. In complex adaptive systems, a system's past forms part of its trajectory and affects its present and future dynamics. Studies of the *longue durée* help us to understand system trajectories and to predict and prepare for potential transitions.

References to the past may be used in different ways. The long-term perspective may be used to refine the conceptual framework of resilience, but the past may also be used as a source of knowledge to further understanding of present societies and in defining the modalities of response to major changes (learning from the past). Work on resilience brings together the historical sciences and the sciences of the present on an equal footing, rather than as auxiliary contributors to discipline-specific work (Robert 2017). Archeology and archeogeography have an important role to play, casting light on the material elements produced by societies and providing tools for trans-disciplinary work (environmental analysis, aerial imaging, etc.). The conceptual framework of resilience is particularly valuable in archeogeography, notably in its insistence on the idea that memory, permanently imprinted upon markers in a landscape, plays an important role in societal resilience.

The conceptual framework of resilience has a powerful interdisciplinary reach, drawing on more general theories common to multiple scientific communities, including notions of complexity, chaos, non-linear system dynamics and self-organization. In ecological resilience, change is considered as an inherent part of a system. Resilience is not the ability to return to a past state, nor is it a state of inertia. One important feature of the ecological understanding of resilience is the idea that a structure can only survive in the long-term if it undergoes transformation(s). The form or structure of a system is never "finished"; it is the result of a structuring process which develops from, and is maintained by, multiple interactions on different spatial and temporal scales. This vision echoes the approach to the dynamics of plot and road systems used in archeogeography since the 1990s (section 3.1).

Nevertheless, adoption of the conceptual framework of ecological resilience by French academics in the fields of geography and the historical sciences has been slow and limited. Attempts to introduce the concept in geography and archeogeography in the early 2000s were only partially successful, barely touching notions such as the adaptive cycle, panarchy and, to a lesser extent, attractors and basins of attraction. Concrete applications have been extremely limited, hindered by a pervasive critical view of resilience as a discursive construct. Generally speaking, applications of resilience to the social domain and an increased use of long-term perspectives favor the adoption of a framework originally developed for ecosystems. Human societies, characterized by an ability to learn and to innovate, exhibit other adaptation mechanisms. The analysis of cultural landscapes is one area within the social sciences in which the conceptual framework of ecological resilience is currently being tested. Archeogeographical analysis of landscape forms, drawing on the theory of complex systems, may provide fertile ground for further testing of this framework.

PART 3

# Synthesis: Landscape as a Resilient Social-ecological System

# Introduction to Part 3

In the first part of this book, we showed that the dominant approach to landscape dynamics prior to the 1990s was based on a separation of past and present forms, a separation which itself had its roots in the nature/culture dichotomy. The idea that a social order may be associated with a spatial order may result in two approaches: either a valorization of past forms, intended to protect them from the transformations of the present, or the consignment of these past forms to an "obsolete" past in order to accelerate societal modernization. In every case, the chosen models – drawing on disciplines from art history to technology – were designed as an expression of the dominance of human reason over nature. Landscape forms were projected and studied with no regard for the geohistorical trajectory in which they exist (Chapters 1 and 2). In the 1990s, a growing awareness of the existence of non-linear temporalities in plot and road networks led archeogeographers to abandon the semiological interpretation of forms, which relies on the traditional past/present, nature/culture divisions. In archeogeography, forms are studied within their geographical context, using conceptual frameworks to examine objects as part of a historical and geographical trajectory (Chapter 3).

Ecological resilience, which draws on the broader theories of complex systems and self-organization, may provide a heuristic framework for current landscape studies (Chapter 4). English-speaking archeologists participated in the development of this framework from the 1990s on, notably via the Resilience Alliance network, demonstrating the value of historical examples in establishing the role of different temporal and spatial scales in long-term resilience (Chapter 5). In our view, archeogeography provides fertile ground for testing the conceptual framework of resilience; the theoretical constructs of complex systems and self-organization have been used in the field since the late 1990s, and the role of interactions between spatial and temporal levels is already clearly established. Archeogeographers also led the way in recognizing the potential of the theoretical framework of resilience in the early 2000s (section 5.1.4). The tools developed within the conceptual framework of ecological resilience for understanding the articulation between persistence and change are echoed in some of the ideas we put forward for revising and improving approaches to morphological analysis.

The theme of cultural landscapes provides a particularly promising basis for transdisciplinary research, resulting from a close association between societies and environments, and between multiple temporal and spatial scales. Furthermore, shifting the focus of research away from disaster management would enable researchers to focus on resilience as a systemic property instead of a discursive concept. Following the ecologist A. Kinzig, the example of landscapes may be used "to better understand how social-ecological systems do or can work, rather than making a normative articulation of how they should work" (Kinzig 2012, p. 325).

An understanding of current debates surrounding the notion of resilience in geography (section 5.2) is essential in order to fully grasp the implications of its use in the present academic climate and to avoid falling into the same traps encountered in the past. There is a risk of assigning excess value to old plans or layouts, notably planned arrangements, so that the production of forms in a landscape is devoted to helping past forms to resist the present. The use of "solidly-built" forms, made to last (such as a centuriation, which "braves the passage of time unscathed", or a constructed road, see sections 1.1 and 1.2) is broadly equivalent to the role of technology in engineering resilience. In culturalist urbanism, as in engineering resilience, an initial state is established as a norm, resulting in a certain ordering of forms in a landscape. The theoretical aim would be to re-establish this norm following perturbations. This perspective is based on the idea of a return to a unique, stable state of equilibrium (on the subject of stability, see section 4.2). However, in this context, the value of the norm is never questioned, and there is an inherent idea that time is reversible, meaning that the trajectory of studied systems is essentially ignored. Time, as an agent of system dynamics, is absent from engineering resilience because of this notion of reversibility. Engineering resilience draws on the nature/culture dichotomy, ignoring the social dimension of catastrophes, seen solely as external perturbations inflicted by nature. It is this form of resilience that has been met with such extensive criticism in the field of geography (section 5.2.5).

In ecological resilience, on the other hand, the relation between societies and environments plays a key role, with a focus on integrated social-ecological systems. According to the ecological understanding of resilience, time is an agent in the transformation of systems, which possess their own dynamics leading to phases of decline, reorganization, etc. (adaptive cycles, panarchy: see sections 4.3 and 4.4). Given the emphasis on the trajectory of social-ecological systems, ecological resilience is a particularly promising concept for use in morphological analysis, within the broader field of archeogeography. Work carried out on applying complex system theory to spatial systems, in the field of geography (section 5.1.2), creates a bridge between archeogeography and the conceptual framework of ecological resilience. The following pages provide clear definitions of concepts that may be used to articulate morphological landscape analysis, as it is understood by present-day archeogeographers, with reflection on the subject of resilient social-ecological systems. We compare the concepts used by archeogeographers, geographers and ecologists, studied throughout this book, showing the many and varied correspondences that exist between these domains, concealed beneath differing formulations and terminologies.

# 7

# Landscape: An Integrated System of Societies and Environments

Archeogeographers and ecologists appear to have had a shared object of study as early as the 1990s in the form of integrated systems of societies and environments. This concept moves beyond the traditional nature/culture dichotomy, viewing certain spatial systems as complex, integrated systems in which man and nature form part of a single dynamic system, connected by feedback loops (Table 7.1). This focus on the relations between nature and culture is present in the notion of the **geosystem**[1], introduced by G. Bertrand and taken up by many archeogeographers in the course of the 1990s, and in that of the **anthroposystem** (*anthroposystème*), a term coined by French researchers in the early 2000s (section 3.1.2). It is also key to the definition of **social-ecological systems**, which form the basis for research on resilience in the English-speaking world (see Introduction to Part 2). The ecologist Y. Gunnell considers the anthroposystem and social-ecological system concepts to be similar, as human and nature form part of the same dynamic system (Gunnell 2009). Kirchhoff *et al.* also argue that the notion of **cultural landscape** is similar to that of social-ecological systems (section 6.3). Nevertheless, the systemic dimension is less readily apparent here than in the other notions.

The **morphosystem**, a concept used in archeogeography (section 3.1.2), offers a means of studying the part of these integrated nature–society systems expressed through the production of planimetric forms that are sufficiently durable to be observed. Seen as a morphosystem, a landscape is thus a meeting-point between archeology and geography, with which it shares many traditional objects of study:

1 The words given in bold correspond to entries in the tables in Chapters 7 and 8, which provide a summary of the main definitions discussed in this book. Words are cited according to the order of appearance in Part 3, and are given in English for ease of reading. In cases where the original text is in French, this is cited in the footnotes.

line networks (roads, plot boundaries, etc.), networks of points (establishments) and "patchworks" (parcels, territorial boundaries, etc.).

| Terms used | Points of discussion | Fields of research |
|---|---|---|
| **Geosystem or GTP (*géosystème, territoire, paysage*)** | "Enables us to grasp all of the relations which exist in a given portion of space, from abiotic potential to biological exploitation and anthropic use"[2] (Bertrand 2002, p. 132). | Geography<br>Archeology |
| **Anthroposystem** | "An interactive system featuring two groups made up of one or more sociosystems and one or more natural and/or artificialized ecosystems, implanted in a given geographical space and evolving over time"[3] (Muxart *et al.* eds 2003, p. 18). | Archeology<br>Geography |
| **Social-ecological system (SES)** | "An SES is an ecological system intricately linked with and affected by one or more social systems" (Anderies 2004). | Ecology<br>Geography |

**Table 7.1.** *Ideas used in expressing and understanding integrated nature–society systems (Robert, 2019)*

| Terms used | Points of discussion | Fields of research |
|---|---|---|
| **Cultural landscape** | "Combined works of nature and humankind" (UNESCO 1992). | Historical ecology<br>Geography |
| **Morphosystem** | Study of the planimetric forms of a landscape, "be they of organic or anthropic origin"[4], interacting with the ecosystem and the sociosystem (Chouquer 1997, p. 18). | Archeogeography |

**Table 7.2.** *Ideas used in expressing and understanding the material translation of integrated nature–society systems (Robert, 2019)*

2 *Permet de saisir l'ensemble des relations qui existent sur une portion d'espace donnée entre un potentiel abiotique, une exploitation biologique et une utilisation anthropique.*

3 *Système interactif entre deux ensembles constitués par un (ou des) sociosystème(s) et un (ou des) écosystème(s) naturel(s) et/ou artificialisé(s) s'inscrivant dans un espace géographique donné et évoluant dans le temps.*

4 *...qu'elles soient d'origine organique ou anthropique.*

The morphosystem concept places a particular emphasis on the spatial dimension of the integrated nature–society system (Table 7.1). The role assigned to flows in the resilience of forms in the archeogeographical approach (sections 3.2.2 and 3.5.4) highlights the potential articulations between archeogeography, geography and ecology. Certain elements, such as river corridors, have already been the subject of transdisciplinary studies involving researchers from these three fields (section 3.3.1) (Pinoteau and Di Pietro 2003); these elements may be interpreted as social-ecological systems.

8

# Landscape as a Complex Adaptive System

Integrated systems of societies and environments can be viewed as complex adaptive systems, described by Gunderson and Holling (2002) as systems "in which components adapt and change with their environments, leading to unpredictable outcomes" (p. 3). Their study corresponds to the more general theoretical frameworks of complex systems and self-organization, which are also used in archeogeography and geography, providing a bridge and an area of common interest between these disciplines and ecology. However, although the dynamics of integrated society–environment systems seem to be understood in the same way, these convergences are partly obscured by differences in terminology.

## 8.1. Landscape emergence as a system: key concepts

To the best of our knowledge, no precise description of the conditions of appearance of social-ecological systems has been established with regard to ecological resilience; for example, the Resilience Alliance glossary makes no mention of the notion of emergence. In archeogeography and geography, on the other hand, the emergence of systems is a particularly important object of study, and one which may be approached in a variety of ways (Table 8.1).

As we have seen, the term **morphogenesis** is used in archeogeography to describe the creation of a new form making up a whole. The recurrence of different morphological principles (isoclination, isotopy, etc.) reveals regularities in landscape forms, making it possible to differentiate between an organization and its environment; these principles also suggest that the structure of a self-organized network is the result of interconnections between different elements (section 3.4). In our view, the notion of morphogenesis, used in archeogeography, corresponds to that of **systemogenesis**, developed by the geographer F. Durand-Dastès (2004) to describe "the period in which the relationships between the elements which make up a system are established"[1] (p. 24). Another geographer, Denise Pumain, has highlighted the notion of **emergence** or the "emergent structure" (*structure émergente*), a term borrowed from physics and the life sciences (section 5.1.2.1).

| Terms used | Points for discussion | Fields of research |
|---|---|---|
| **Morphogenesis** | The appearance of "a reality which is topologically, structurally and qualitatively new in space and time"[2] (Morin 1977, p. 115).<br>"The study of the modalities of the creation and evolution of forms"[3] (Chouquer 2000, p. 188). | Archeogeography |
| **Systemogenesis** | "The period in which the relationships between the elements which make up a system are established"[4] (Durand-Dastès 2004, p. 24).<br>"The formation of a system, i.e. the appearance of strong organizations in a situation where there were not previously any looping interactions between present elements"[5] (Durand-Dastès 2010, p. 16). | Geography |
| **Emergence** | "In relation to the pre-existing system, emergence may consist of the creation of a new category of elements, or the appearance of a new structure, or attributes, or in the appearance of a new structure, even a new system, or in the invention of new rules by the protagonists of a simulation game"[6] (Pumain 2003, § 19). | Geography |

**Table 8.1.** *Ideas used in expressing and understanding the appearance of new spatial and landscape organizations (Robert 2019)*

1 *...période où se nouent les relations entre les éléments qui constitueront le système.*

2 *...une réalité topologiquement, structurellement, qualitativement nouvelle dans l'espace et le temps.*

3 *Étude des modalités de création et d'évolution des formes.*

4 See above.

5 *...la formation d'un système, c'est-à-dire l'apparition d'organisations fortes dans une situation où il n'y avait pas eu jusque-là d'interactions en boucles entre les éléments présents.*

6 *Par rapport au système préexistant, l'émergence peut consister dans la création d'une nouvelle catégorie d'éléments, ou d'attributs, ou dans l'apparition d'une nouvelle structure, voire d'un nouveau système, ou dans l'invention de nouvelles règles par les protagonistes d'un jeu de simulation.*

The phase in which connections between establishments and their associated plot layouts, alongside structural landscape elements such as road or hydrosedimentary networks, are established – developments observed in the late Iron Age and the 1st century AD in France – may be seen to represent the emergence of self-organized, temporally resilient systems (section 3.5.2).

## 8.2. Maintenance and reproduction of landscapes as systems: key concepts

In ecology, the notion of resilience is used to describe the continued existence of ecosystems in spite of the effects of perturbations. In **ecological resilience**, systems are not considered to be in a state of equilibrium; indeed, perturbations play an important part in their evolution. Resilience can be calculated in terms of the size of a basin of attraction (Holling 1973, 1978, p. 11; Scheffer *et al.* 2003, p. 650). In **engineering resilience**, systems do possess an equilibrium state, and resilience is calculated in terms of the time taken to re-establish this equilibrium state following a perturbation (Pimm 1984).

In geography, a variety of terms, such as "inertia", "legacy", "inheritance", "reproduction" and "spatial memory" have been used with respect to the maintenance of a system over time (section 5.1.2.1). **Resilience** came into use as a systemic property from the 2000s, but its adoption was very limited until the 2010s (sections 5.1.3, 5.1.4 and 5.2). At the time of writing, engineering resilience is subject of extensive criticism with regard to its use as a discursive construct in the context of disaster management (section 5.2.4).

In archeogeography, resilience as a systemic property made its first appearance in the early 2000s (Chouquer 2000; Robert 2003a), but the dynamics of plot layouts and road networks were essentially defined using the concept of **morphostasis** (Marchand 2000, 2003), then **transmission** (and the idea of **transformission**: Chouquer 2003b) up until the 2010s (section 3.5.3). The notion of transmission is currently giving way to that of resilience. **The term "resilience" may thus form part of a shared terminology for archeogeography, geography and ecology, expressing the way in which an existing spatial organization maintains its form and is reproduced over time**. The interest of resilience (in the ecological sense) is that it draws attention to the systemic dimension, which is less evident in the term "transmission". Resilience may notably be used to define the process by which forms are maintained at the macroscopic level.

| Terms used | Points for discussion | Fields of research |
|---|---|---|
| **Ecological resilience** | "…a property that allows a system to absorb and *utilize* (or even benefit from) change" (Holling 1978, p. 11).<br>"Resilience is the capacity of a system to absorb disturbance and reorganize while undergoing change so as to still retain essentially the same function, structure, identity, and feedbacks" – or, in other terms, to remain within the same basin of attraction (Walker *et al.* 2004, p. 6). | Ecology<br>Geography<br>Archeology (En.)<br>Archeogeography |
| **Engineering resilience** | Calculation of the speed at which the variables characterizing a complex, stable ecosystem return to a state of equilibrium following a perturbation, expressed in units of time (Pimm 1984, p. 322). | Ecology<br>Geography |
| **Resilience applied to spatial systems** | "…the measure of a system's, or part of a system's capacity to absorb and recover from the occurrence of a hazardous event" (Timmerman 1981, p. 21).<br>"[a] *capacity to adapt* to external change, but also to produce or adopt technical and cultural innovations, ensuring the perennity of a social group in its environment, consisting of both the natural milieu and other, competing social groups"[7] (*Archaeomedes* 1998, p. 14, italics original).<br>"The capacity of a system to integrate a perturbation into its functioning without altering its qualitative structure"[8] (Aschan-Leygonie 2000, p. 64).<br>"…the ability of groups or communities to cope with external stresses and disturbances as a result of social, political and environmental change" (Adger 2000, p. 347). | Archeology<br>Geography<br>Archeogeography |
| **Morphostasis** | The way in which an organization "maintains the permanence of the system in its form (*Gestalt*), existence and identity"[9] (Morin 1977, p. 131).<br>Dynamic processes responsible for the formation and reproduction of protohistoric and ancient plot layouts (Marchand 2000). | Archeogeography |

7 … faculté d'adaptation *à des changements externes mais aussi à la production ou à l'adoption d'innovations techniques et culturelles, qui permet d'assurer la pérennité d'un groupe social dans son environnement, lequel est constitué aussi bien par le milieu naturel que par d'autres groupes sociaux concurrents.*

8 *Capacité d'un système à intégrer une perturbation dans son fonctionnement sans changer de structure qualitative.*

9 … *maintient la permanence du système dans sa forme (*Gestalt*), son existence, son identité.*

| Terms used | Points for discussion | Fields of research |
|---|---|---|
| **Transmission-transformission** | Neologism formed from the words "transformation" and "transmission": "A dual action of the transformation of geographical realities and the transmission of some characteristics of these realities over time, creating an impression of permanence in the form" (Chouquer 2003b, p. 303)[10].<br>Expresses the fact that "the system does not reproduce itself in an identical form; it lasts by self-transforming"[11] (Chouquer 2000, pp. 109–110). | Archeogeography |
| **Resilience applied to landscapes** | The capacity of plot systems and road networks to retain their identity, using conflicts of forms and drawing on the interactions between forms on different levels (flows, ground footprints, shapes), permitting an articulation of global and local scales (Robert 2003b, p. 457 and 470).<br>"The capacity of a set of forms (a network of lines, points, etc.) to maintain its structure despite changes in social formations and the transformations resulting from these changes. Resilience is a principle of stability in change, up to a point of rupture or division"[12] (Chouquer 2003a, p. 302).<br>"Resilience may be defined as a landscape's capacity to retain its identity, i.e. to maintain the relations, morphogens and morphological principles which characterized it on a macroscopic level, by absorbing, or even using, the changes which affect it. Resilience can be used to describe the point to which a landscape can self-organize and is able to adapt" (Robert: see below, section 8.2). | Archeogeography |

**Table 8.2.** *Ideas used in expressing and understanding the processes by which social-ecological systems, spatial systems and landscape forms are reproduced (Robert 2019)*

10 *Double action de transformation dans le temps des réalités géographiques et de transmission de certains caractères de ces réalités donnant l'impression d'une pérennité de la forme.*

11 *...le système ne se répète pas à l'identique ; il se pérennise en se transformant.*

12 *...l'aptitude d'un ensemble de formes (un réseau de lignes, de points, etc.) à maintenir sa structure alors que les formations sociales ont changé et le transforment. La résilience est un principe de stabilité dans le changement jusqu'à un point de rupture ou bifurcation.*

We have chosen to focus on those definitions of resilience which appear best-suited to the study of landscape forms in archeogeography, particularly those which highlight the capacity of a system to absorb, and derive benefit from, perturbations in order to adapt while maintaining their structure. The definition of resilience which we propose here draws on those established by Holling (1973, 1978) and that found on the Resilience Alliance website[13], adapted for application to landscape. **Resilience may be defined as a landscape's capacity to retain its identity, i.e. to maintain the relations, morphogens and morphological principles which characterized it on a macroscopic level, by absorbing, or even using, the changes which affect it. Resilience can be used to describe the point to which a landscape can self-organize and is able to adapt.**

This definition draws attention to the absorption of change, but also to its exploitation. In archeogeography, the asynchronous transformation of the different levels inherent in landscape forms (flows, ground footprints, shapes) are essential for long-term resilience (section 3.5.5). Change must thus be considered as a corollary to survival, not simply as an external perturbation. Change forms part of a system's dynamics, notably in terms of its implications for learning and adaptation (Table 8.2).

## 8.3. Temporalities of resilience in a landscape: key concepts

The greatest number of terms discussed here relate to the continued existence of past forms in the present (Table 8.3). In the late 19th century, ideas such as the **palimpsest**, in history, the **principle of continuity** or the **speaking masses** (*masse parlante*) in the work of F. de Saussure (section 1.2.2) and **natural sentiment** (*sentiment naturel*) for P. Lavedan (section 1.2.2) all expressed the idea that the present depends, to a great extent, on elements organized in the past. The same idea is partially expressed by the notion of **inertia**, borrowed from the field of mechanics. **Diachrony**, a term coined by F. de Saussure, was used to describe a "sliding" "down the slope of time" (Braudel 1958, p. 739), an image used to interpret the evolution of states over time. The concept of diachrony has been widely used across the social sciences.

From the 1990s on, other terms, such as **heritage**, **legacy or inheritance**, "received space", or the different types of memory put forward by F. Durand-Dastès (**inscribed memory**, **spatial memory**, **memory space**: *mémoire inscrite, mémoire spatiale, espace mémoire*) (section 5.1.2.1) were used in geography to denote the process by which certain material organizations may persist in a place despite the

13 Resilience Alliance–Glossary. https://www.resalliance.org/glossary, accessed September 15, 2017.

disappearance of the original causes (Table 8.4). The notion of **diachrony** has been used to describe this phenomenon in geography, archeogeography and history. However, it is not always evident whether diachrony is being used to characterize the phenomenon itself, or if it describes the perspective used by the researcher in describing the phenomenon. In archeogeography, G. Chouquer used diachrony as one of the internal morphological time modes used to describe the permanence of landscape forms (section 3.6.1); nevertheless, he indicated that the concept was "fuzzy in terms of the dynamics of forms" and thus "too general to be operational"[14] (Chouquer 2000, p. 126). While the notions of diachrony, inertia, heritage and memory are clearly defined as elements of system dynamics, Durand-Dastès also used these terms in their more general sense, meaning that they cannot be assumed to relate directly to the field of complex systems.

| Terms used | Points for discussion | Fields of research |
|---|---|---|
| **Palimpsest (Late 19th century, 1930s)** | "...from the ordnance map (that marvellous palimpsest, which under Dr Meitzen's guidance we are beginning to decipher)" (Maitland 1987, p. 15-16).<br>"And yet, the primitive core of the population can still be discovered"[15] (Vidal de La Blache 2015, § 17).<br>"Field layout is the book in which rural societies inscribed, line by line, the vicissitudes of their past" (Bloch 1934, p. 483).<br>"That these ancient, favored lines have survived beyond orogeny, like the hidden letters of a palimpsest, is clear..."[16] (Demangeot 1943, p. 571). | History<br>Geography<br>Archeology |
| **Principle of continuity**<br>**Speaking masses** | Idea that the persistence of initial matter ensures the continuity of the sign over time (de Saussure 1995a, p. 119).<br>The social mass is "naturally inert" and acts as a powerful conserving factor, via the "principle of continuity, counterbalancing freedom"[17] (de Saussure 1995a, p. 112). | Linguistics |

14 *...flou en matière de dynamique des formes [...] trop général pour être opératoire.*

15 *Le noyau primitif du peuplement se laisse encore discerner.*

16 *...survie étonnante de ces lignes privilégiées anciennes par-delà les orogenèses, telles les lettres cachées d'un palimpseste...*

17 *...principe de continuité, qui annule la liberté.*

| Terms used | Points for discussion | Fields of research |
|---|---|---|
| **Natural sentiment** | Phenomenon by which "an owner will spontaneously rebuild his dwelling where it was before"[18] (Lavedan 1926a, pp. 92–93). | Architecture |
| **Inertia** | "The resistance of a body to movement, due to its mass" (Larousse 1922, p. 1183)[19].<br>"The process by which an activity continues in a place after its initial causes have disappeared"[20] (Durand-Dastès 1984, p. 13). | Geography |
| **Diachrony** | The area of linguistics which describe successive states in a language, and the phenomena which cause a language to shift from one state to another (de Saussure 1995b).<br>"Similarly, a descent down the slope of time can only be envisaged in the form of multiple descents down the various and innumerable rivers of time"[21] (Braudel 1958, p. 739). | Linguistics<br>History |

**Table 8.3.** *Ideas used in expressing and understanding the temporalities of the permanence of forms in a landscape prior to the 1970s–1980s (Robert 2019)*

The notions of **path dependence** and **hysteresis** may be used to describe the temporalities of resilience of past forms in the present, assimilating them to complex systems with multiple stable states. The current state of a system is explained by its trajectory or history; the path taken has an impact on both present and future dynamics. Hysteresis and path dependence are used to describe this "system memory" (section 4.5.2). In the social sciences, hysteresis is used to express a "cause-effect delay" (*retard de l'effet sur la cause*) (Baudelle and Régnault 2004, p. 65)[22]. Spatial

18 *...le geste spontané du propriétaire est de rebâtir sa demeure où elle était.*

19 *Résistance que les corps, grâce à leur masse, opposent au mouvement.*

20 *Processus selon lequel une activité se maintient dans un lieu lorsque ses causes initiales ont disparu.*

21 *De même, une descente selon la pente du temps n'est pensable que sous la forme d'une multiplicité de descentes, selon les diverses et innombrables rivières du temps.*

22 In sociology, for example, Pierre Bourdieu describes a "hysteresis of habitus" (*hystérésis de l'habitus*): the fact that individuals retain the dispositions acquired during their socialization, even if they move into a different social space (change in status, social position, or evolution of the social space (Fabiani 2016).

temporalities are never completely synchronized with social or economic temporalities due to a "response time" or "reaction time", defined in terms of hysteresis; this explains the existence of a "form lag" or "delay" (section 5.1.2.1).

The term **hysterechrony** has been put forward in archeogeography as a more precise alternative to diachrony for explaining delays between the installation of landscape elements at a time t and the observation of their effects, which may occur over a much longer period via principles of isoclination, isotopy and iso-axiality (Chouquer 2000). This temporality makes it possible to explain the long-term morphogenetic effect of certain elements (morphogens) in reproducing a certain number of morphological principles (section 3.5). The notions of hysterochrony and hysteresis could be combined, under the latter term, unifying the vocabulary used by ecologists, geographers and archeogeographers.

The idea of **uchrony** also seems to describe the dynamics of social-ecological systems and of complex systems in general. It indicates the existence of several possible pathways and highlights the importance of a system's trajectory. In an article published in 2000, C. Aschan-Leygonie used the image of an "epigenetic landscape", developed by the biologist Conrad Hal Waddington, to illustrate the different possible trajectories of a system within a single domain of attraction[23]. In a complex system, an infinitesimal difference in initial system conditions may draw the system into a different basin of attraction (section 4.5.1). In uchrony, changes in certain historical events, even centuries distant, could result in entirely different historical scenarios (section 3.6.1). Applying uchrony to archeogeography, any landscape element imprinted in a space at a time t has a potential role to play in the trajectory of the morphological system, even if only a trace remains. In the material world, even an abandoned human construction may continue to affect the landscape, as we see in the case of archeological traces observed through aerial photography, landscape forms and land-use legacies (sections 2.1.2 and 6.3.1). Forms imprinted in a space at a time t, in a particular historical context, have the potential to be re-used and to affect the trajectory of a social-ecological system, even if they have long ceased to be active or even, in some cases, visible. These non-linear temporalities are explained in part by the articulations between permanence and change within the concept of resilience.

23 Image proposed by C.H. Waddington in 1957 to illustrate cellular differentiation based on the notion of sensitivity to initial conditions (Bourgine and Lesne 2006), reproduced in Aschan-Leygonie (2000, p. 66).

| Terms used | Points for discussion | Fields of research |
|---|---|---|
| **Heritage, inheritance, legacy** | "localization process in which concrete traces left by the past are one of the causes at work in the present"[24] (Durand-Dastès 1984, p. 13).<br>Phenomenon according to which "once produced, any spatial system has an effect on the economic and social system which designed it, at least in terms of the necessary difference between the spatial projects of human groups and their realization"[25] (Baudelle and Régnault 2004, p. 64). | Geography<br>Archeogeography |
| **Received space** | "Thus, every generation has a "received space" […] a space which possesses characteristics which are not dependent on itself"[26] (Durand-Dastès 1986, p. 116). | Geography |
| **Inscribed memory**<br>**Spatial memory**<br>**Spatialized memory** | "The spatial inscription of development and transformation actions upon the earth, transmitted by the presence of concrete, material phenomena"[27] (Durand-Dastès 1991, p. 149). | Geography |
| **Diachrony** | Area of linguistics describing the successive states of a language and the phenomenon which cause a language to move from one state to another (de Saussure 1995b).<br>"Forms structuring a space outside of the period in which they were initiated, providing a permanent framework for identical functions"[28] (Chouquer 2000, p. 126). | Linguistics<br>Archeogeography |
| | "For conditions in which alternative equilibria exist, the initial state (i.e. place in the landscape) determines the equilibrium to which the system will settle" (Scheffer and Carpenter 2003). | |

24 *...processus de localisation dans lequel les traces concrètes laissées par le passé sont une des causes qui interviennent dans le présent.*

25 *...une fois produit, tout système spatial agit en retour sur le système économique et social qui l'a conçu, ne serait-ce qu'en raison du nécessaire décalage qui sépare les projets spatiaux des groupes humains et leur réalisation.*

26 *Toute génération recevrait ainsi un "espace reçu" […] espace qui a des caractères qui ne dépendent pas d'elle.*

27 *...l'inscription spatiale des actions d'aménagement et de transformation de la terre ; elle est transmise par la présence de phénomènes concrets et matériels.*

28 *Formes structurant l'espace au-delà de la période qui les a initiées et servant de cadre permanent à des fonctions identiques.*

| Terms used | Points for discussion | Fields of research |
|---|---|---|
| **Path dependence**<br>**Trajectory dependence**<br>**Historical sequence** | "Systems with alternative stable states may display hysteresis (or path-dependence, where the state transition function depends not only on the current state but on the history of previous states)..." (Halvorsen and Layton 2015, p. 176). | Ecology<br>Geography |
| **Hysteresis** | "Cause-effect delay" (Baudelle and Régnault 2004, p. 65).<br>"To induce a switch back to the original valley, it is not sufficient to restore the environmental conditions present before the collapse. Instead, one needs to go back beyond another bifurcation point (F1), where the system shifts back. The difference between forward and backward switches is known as hysteresis" (Scheffer *et al.* 2003, p. 650). | Geography<br>Ecology |
| **Hysterechrony** | "Hysterechrony, or a temporal mismatch between a form and social formations with the functions which they imply, describes the spatio-temporal modality of maladjustment in the morphological system. It results in morphological hysteresis, an observable lag between a 'cause' and its effect"[29] (Chouquer 2000, p. 125).<br>"Offsets, delays and response times observed in the dynamics of forms"[30] (Chouquer 2003b, p. 302). | Archeogeography |
| **Uchrony** | "A potential which a later, new social fact (any development or transformation of the space) may draw upon at a later, unknown date in the history of the site"[31] (Chouquer 2000, p. 126). | Archeogeography |

**Table 8.4.** *Ideas used in expressing and understanding the temporalities of permanence in spatial systems, social-ecological systems and landscape forms from the 1990s onward (Robert 2019)*

29 *L'hystéréchronie, ou décalage dans le temps de la forme par rapport aux formations sociales et aux fonctions qu'elles induisent, décrit la modalité spatio-temporelle des inadaptations du système morphologique. Elle produit l'hystérésis morphologique ou temps de latence constaté entre une "cause" et son effet.*

30 *Décalages, retards et temps de réponse qu'on observe dans la dynamique des formes.*

31 *Un potentiel que des faits (tout aménagement et toute transformation de l'espace) ultérieurs peuvent faire rejouer.*

| Terms used | Points for discussion | Fields of research |
|---|---|---|
| **Bifurcation** | "Critical point after which a new state becomes possible"[32] (Prigogine and Stengers 1990, p. 229).<br>"Resilience is a principle of stability in change, up to a point of rupture or division"[33] (Chouquer 2003a, p. 302).<br>"A change in the very nature of the attractor towards which the system tends"[34] (Sanders *et al.* 2020a, p. 135). | Archeogeography<br>Geography |
| **Transition** | "In the case where the system changes in a radical manner, i.e. disappears and/or is transformed into another, qualitatively different system, the initial 'basin of attraction' is abandoned in favor of a new attractor [...]. When this process occurs over a certain period of time, this constitutes a transition from one system to another"[35] (Sanders *et al.* 2020a, p. 139). | Geography<br>Archeogeography<br>Archeology |
| **Regime shift** | "Sudden shifts in ecosystems, whereby a threshold is passed and the core functions, structure, and processes of the new regime are fundamentally different from the previous regime"[36] (Resilience Alliance 2019). | Ecology |
| **Tipping point** | "Tipping point (i.e. a catastrophic bifurcation), where a minor trigger can invoke a self-propagating shift to a contrasting state" (Scheffer *et al.* 2012, p. 344). | Ecology |
| **Creative destruction**<br>**Release phase**<br>**Ω phase** | "Phase of release or "creative destruction", when tightly bound resources become increasingly fragile and then are suddenly released" (Resilience Alliance 2019)[37]. | Ecological resilience |

**Table 8.5.** *Ideas used in expressing and understanding transitions in social-ecological systems, spatial systems and landscape forms (Robert 2019)*

## 8.4. Transitions in landscapes: key concepts

Archeogeographical studies tend to emphasize the resilience of forms, as these indicate the presence of ancient organizations, which can still be seen from modern and contemporary planimetric sources. Combining these findings with archeological data, it

32 *Point critique à partir duquel un nouvel état devient possible.*

33 *La résilience est un principe de stabilité dans le changement jusqu'à un point de rupture ou bifurcation.*

34 *Un changement dans la nature même de l'attracteur vers lequel évolue le système.*

35 *Dans le cas où le système change de manière radicale, c'est-à-dire disparaît et/ou se transforme en un autre système qualitativement différent, le "bassin d'attraction" initial est abandonné au profit d'un nouvel attracteur [...]. Quand ce processus s'inscrit dans une certaine durée, il y a transition d'un système vers un autre.*

36 *Resilience Alliance–Glossary*. https://www.resalliance.org/glossary, accessed May 28, 2019.

37 *Resilience Alliance–Glossary*. https://www.resalliance.org/glossary, accessed May 28, 2019.

is also possible to study cases where plot or road networks appear to have been used over much shorter periods. In archeogeography, the term **bifurcation**, borrowed from the language of the complex sciences and of self-organization (Prigogine and Stengers 1990, p. 229), is widely used to describe the "breaking point" marking the disappearance or appearance of a new network (Chouquer 2000, pp. 123–124) (Table 8.5).

Geographers use the terms "bifurcation" and "**transition**" to describe the passage from one system to another, which is qualitatively different. L. Sanders differentiates between the two notions based on the duration of the passage phenomenon:

> In the case where the system changes in a radical manner, i.e. disappears and/or is transformed into another, qualitatively different system, the initial 'basin of attraction' is abandoned in favor of a new attractor […]. When this process occurs over a certain period of time, this constitutes a transition from one system to another and, in cases where the change occurs rapidly with respect to the time scale associated with the phenomenon in question, we use the concept of bifurcation[38]. (Sanders *et al.* 2020a, p. 139)

In ecology, the expression **regime shift** appears to be preferred to the terms "transition" and "bifurcation" (e.g. Scheffer and Carpenter 2003; Kinzig *et al.* 2006; Gunderson *et al.* 2017) (section 4.5.1). The Resilience Alliance glossary features an entry for Regime Shift, but not for transition or bifurcation. Searching for all three terms in articles on the *Ecology and Society* journal website, we find three articles in which "bifurcation" is used, 142 using "transition", and 302 containing the term "regime shift"[39]. The notion of a regime shift is defined as "[A] sudden shift in ecosystems, whereby a threshold is passed and the core functions, structure, and processes of the new regime are fundamentally different from the previous regime"[40]. The expressions "catastrophic shift" (Scheffer *et al.* 2001) and "catastrophic regime shift" (Scheffer *et al.* 2003) have been used to describe relatively rapid passages from one basin of attraction to each other, and may thus be assimilated to the notion of bifurcation. In archeogeography, we believe that the term "transition" is most appropriate, given that the transformations studied in this discipline generally occur over a certain period of time.

---

38 *Dans le cas où le système change de manière radicale, c'est-à-dire disparaît et/ou se transforme en un autre système qualitativement différent, le "bassin d'attraction" initial est abandonné au profit d'un nouvel attracteur […]. Quand ce processus s'inscrit dans une certaine durée, il y a transition d'un système vers un autre et, dans les cas où le changement est rapide relativement à l'échelle de temps associée au phénomène étudié, on mobilise le concept de bifurcation.*

39 Search carried out July 3, 2019 at https://www.ecologyandsociety.org.

40 Resilience Alliance–Glossary. https://www.resalliance.org/glossary, accessed May 28, 2019.

The **reorganization (alpha) phase** in C.S. Holling's adaptive cycle model may be interpreted as a regime shift. In this case, the preceding **release or creative destruction phase** can be considered to describe the moment when the system switches to a new basin of attraction, "when tightly bound resources become increasingly fragile and then are suddenly released" (Resilience Alliance 2019). However, it is not clear whether this phase and the subsequent reorganization phase may be interpreted as a true regime shift (section 6.1).

## 8.5. Reorganization in a landscape: key concepts

In archeogeography and geography, reorganization is considered as the appearance of a new system, marked by a new phase of morphogenesis and systemogenesis. The term "emergence" is thus also appropriate (section 5.1.2.1). In Holling's adaptive cycle, the **alpha** or **reorganization phase** of the cycle describes the specific moment in the dynamics of social-ecological systems where the system undergoes reorganization following a phase of disturbance. However, as we have seen, the meaning of this phase may be ambiguous in the absence of a precise definition of the preceding release (Ω) phase. Does the system move into another basin of attraction, or is this a reorganization within the same basin (section 6.1 and Table 8.6)?

In Part 2, we showed that studies of long-term perspectives by archeologists and archeogeographers are useful in precisely characterizing transitions. Some researchers have used the adaptive cycle and panarchy models to describe the dynamics of social-ecological systems. The notions of attractors and basins of attraction seems to be particularly apt for characterizing transitions (sections 4.1 and 4.5). These notions thus provide tools for a more precise characterization of resilience in landscapes, seen as social-ecological systems. In archeogeography, we must define the precise application of these conceptual tools to our particular objects of study.

| Terms used | Points for discussion | Fields of research |
|---|---|---|
| **Emergence** | Defined in Table 8.1. | Geography |
| **Systemogenesis** | Defined in Table 8.1. | Geography |
| **Morphogenesis** | Defined in Table 8.1. | Archeogeography |
| **Alpha/reorganization phase** | "Resilience and recovery are determined by the release and reorganization sequence" (Holling 1992, p. 450).<br>"A phase in which the system reorganizes following some disturbance and a time for innovation and restructuring" (Resilience Alliance 2019). | Ecology |

**Table 8.6.** *Ideas used in expressing and understanding reorganization in spatial systems, landscape forms and social-ecological systems (Robert 2019)*

## 8.6. The articulation between persistence and change in a landscape: key concepts

Ecologists, geographers and archeogeographers have all attempted to describe the articulation between transformation and persistence in system dynamics, drawing on a number of ideas. In ecology, **panarchy** is one of these themes; in our view, the emphasis on the complex relations between transformation and persistence in this model makes it particularly attractive. Panarchy may be articulated with the idea of **asynchrony between flows, ground footprints and shapes**, which we developed in order to explain the resilience of plot layouts and road networks over long periods of time (section 3.5.5).

The focal scale used in archeogeography is that of ground footprints (plot boundaries, roads and paths, etc.). Analysis has shown that these ground footprints are strongly influenced by circulation networks on the level immediately above, and that they are also dependent on shapes observed on the level immediately below (section 3.5.4). **For plot layouts and road networks, circulation or flow may be seen to constitute the large and slow level of the panarchy model**. Certain elements, such as the circulation of water, goods, persons, etc., constitute common morphogens for landscapes across time and across different cultural contexts (for example, see Robert and Sittler, 2016, for the case of water). Circulation or flow must generally be observed on a broader level and exhibits relatively slow rhythms of transformation, as seen in the case of road networks. Shapes, observed at the micro level, generally have a faster rhythm of transformation than routes and road layouts, corresponding to the small and fast level of the panarchy model. Ground footprints represent the intermediate size and speed level. Their temporalities are relatively varied, from slow to fast, and may include phases of reuse, etc. The three levels are in constant interaction; ongoing readjustments of ground footprints and shapes are essential to maintain circulation on a certain level (section 3.5.4).

| | **Small and fast** | **Intermediate size and speed** | **Large and slow** |
|---|---|---|---|
| **Spatial dimension** | Shape | Ground footprint | Circulation/flow |
| **Rhythm of transformation (years)** | <50, 75, 100 (based on 115 states observed in Robert and Verdier 2014) | <50, 50, 200, 300, 500, or >2000, cases of reuse (from a study of 9 routes in the Val-d'Oise: Robert 2003a) | >2,000 (for 6 out of 8 links between ancient cities in the Parisian basin: Robert 2003a) |

**Table 8.7.** *Example of an adaptation of the panarchy model to the three levels of analysis for road networks: flow, ground footprint and shape (Robert 2019)*

The "**remember**" and "**revolt**" arrows in the panarchy model indicate moments where certain levels affect the rhythms of transformation or maintenance of higher or lower levels. The case of the former Gisors-Pontoise road in Marines (section 3.5.5) may thus be reinterpreted using the panarchy model (Figure 8.1). On the macrolevel, the adaptive cycle relates to the maintenance of rapid regional circulation between Paris and Dieppe (passing through Pontoise and Gisors). The existence of a group of several layouts for the Pontoise-Gisors road at Marines during the Middle Ages may be seen as the exploitation (r) phase of the adaptive cycle: the six branches of the regional itinerary cover a vast space, and are connected to several settlement centers (Figure 3.16). These multiple branches mean that if one layout was interrupted or disturbed in any way, many alternatives were still available. Continued circulation resulted in the construction of road elements of increasing technical sophistication, such as the royal road in the 18th century and the *route nationale* road from the 19th century on. From the 18th century, public investment came to focus on a single road tracing: the branch which passed by the chateau at Marines. The development of the royal, then national, roads may be assimilated to the K, or conservation, phase. Circulation was facilitated by these developments, but the route itself became more vulnerable as it came to depend on a single road tracing.

From the 19th century, activities (industrial, trade, hospitals, etc.) came to be increasingly connected to the layout of the royal/national road, resulting in a progressive slowing of circulation. By the end of the 20th century, conflicts of use affected both local populations (safety, pollution, noise, etc.) and regional traffic, hindered by the passage through the center of Marines. The volume covered by the different elements in question (road, buildings, vehicles, etc.) constitutes the scale of the shape. In 2002, county authorities, in collaboration with the town authorities in Marines, agreed to construct a bypass to resolve these problems (revolt arrow). The path taken by the bypass corresponded, in part, to that of the old medieval road from Pontoise to Gisors, which may be represented by a "remember" arrow (Figure 8.1).

The non-linear temporalities at work with respect to the Pontoise–Gisors road can only be understood with reference to other elements, on both higher and lower levels, interacting with the ground footprint. This example also shows that the panarchy model needs to be adapted for each specific case. The revolt arrow may be seen to come from either the local or global level; traffic saturation was harmful to both levels, and the bypass project was coordinated by both regional and local authorities (the *Conseil Départemental* and the local authorities in Marines). Furthermore, the reuse of the inherited layout may be represented as a memory arrow from the intermediate to the macrolevel (Figure 8.1), whereas, in the classic panarchy model, this arrow tends to point from the macrolevel to the microlevel or intermediary levels (Figures 4.6 and 4.7). Broadly speaking, the application of the panarchy approach to concrete cases results in a more complex, more operational model.

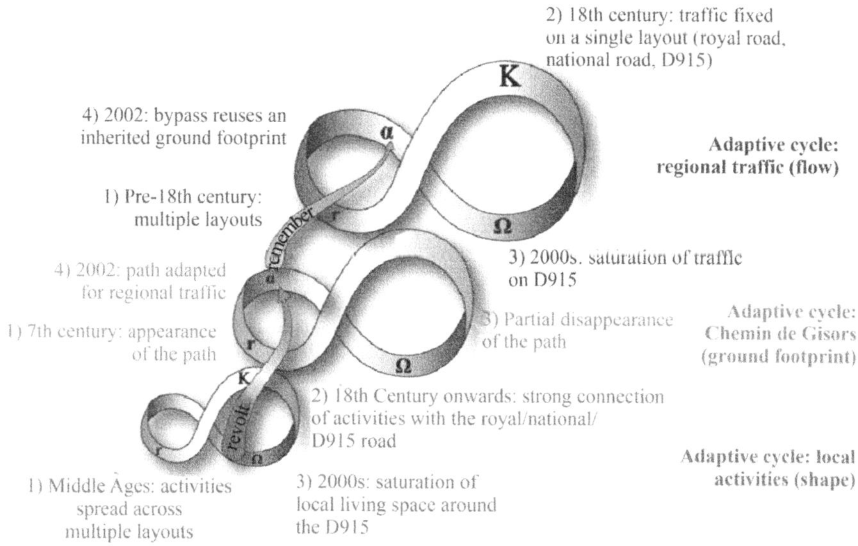


**Figure 8.1.** *Interpretation of the Marines bypass case based on the panarchy model proposed by Gunderson and Holling 2002. Conflicts of use posed a threat to the adaptive cycle of regional traffic (revolt arrow), leading to the creation of a bypass that partly reused the ground footprint of a medieval path (remember arrow) (Robert 2019). For a color version of this figure, see www.iste.co.uk/robert/resilience.zip*

**Hysterechrony and uchrony** are particularly helpful tools for describing the temporalities at work in panarchy:

– Hysterechrony describes lags and response times at the intermediate level with regard to transformations on both the large and slow and small and fast levels. There are many cases in which road layouts are no longer adapted to present-day traffic, which has become increasingly dense over time. Transformations in modes of transport also necessitate a new type of road structure. Revolt arrows appear as attempts to reduce the mismatch between the rhythms of different interacting levels.

– Uchrony, on the other hand, describes remember arrows, in which the ground footprints left by past societies may be reactivated (as in the case of the Marines bypass, the plot boundary at Pierrelatte and the bridge at Cergy: section 3.6.1) (Table 8.8).

| Terms used | Points for discussion | Fields of research |
|---|---|---|
| **Panarchy** | The way in which the states of a system may be influenced by the states and dynamics of systems or sub-systems on higher or lower levels of the focal scale (Walker *et al.* 2004, p. 5 and 7).<br>"A panarchy is a model of linked, hierarchically arranged adaptive cycles that represents the cross-scale dynamic interactions among the levels of a system and considers the interplay between change and persistence" (Resilience Alliance – Glossary)[41]. | Ecology |
| **Asynchrony of flow-ground footprint-shape** | The transformation of ground footprints and shapes at local level, enabling landscape elements to be adapted to new functions, ensures the persistence, at global level, of the circulations (flows) which structure plot and road networks in the long-term (Robert 2003a). | Archeogeography |
| **Revolt arrow** | Events which take place in fast and intermediate cycles can influence slower levels (Holling *et al.* 2002, p. 76).<br>The small, fast level can transform the large, slow level. (Holling 2006, p. 20-21). | Ecology |
| **Remember arrow** | Remember arrows may facilitate reorganization, as the system may draw on the potential accumulated at other levels (Holling *et al.* 2002, p. 76). | Ecology |
| **Hysterechrony** | "Offsets, delays and response times observed in the dynamics of forms"[42] (Chouquer 2003b, p. 302). | Archeogeography |
| **Uchrony** | "Uchrony, or the potentiality of a form in time, describes the spatio-temporal modality which is observed when a structure or a formal element of a structure present in a landscape at a given moment in history creates a potential which a later, new social fact draws upon at a later, unknown date in the history of the site. The modality is then expressed either through isotopy (revival of the form on the same site) or through isoclination (retrieval of the orientation of the form)"[43] (Chouquer 2000, p. 126). | Archeogeography |

**Table 8.8.** *Ideas involved in the articulation of permanence and change in social-ecological systems and landscape forms (Robert 2019)*

41 https://www.resalliance.org/glossary, accessed September 4, 2018.

42 *Décalages, retards et temps de réponse qu'on observe dans la dynamique des formes.*

43 *Uchronie ou potentialité de la forme dans le temps décrit la modalité spatio-temporelle qui se constate lorsqu'une structure ou un élément formel d'une structure imprimée dans le sol à un moment historique donné crée un potentiel qu'un fait social nouveau et ultérieur fait rejouer à un moment imprévu de l'histoire du site. La modalité s'exprime alors soit par isotopie (reprise de la forme au même emplacement), soit par isoclinie (reprise de l'orientation de la forme).*

## 8.7. Temporalities of reorganization in landscape: key concepts

The **palimpsest**, in the sense used from the second half of the 20th century on (in terms of the erasure and replacement of old writing with new) provided a means of expressing the replacement of certain landscape organizations with others (section 2.2.1). In archeogeography, transitions are marked by the disappearance of the morphological characteristics that indicate the reproduction of plot and road networks over time. The morphological temporality used to indicate a "rupture of form in time" is known as **synchrony**, and relates to synchronization between forms and social productions. In the wake of the Second World War, for example, consolidation programs aimed to bring plot and road networks into line with the societal aspirations of the post-war period (section 2.2.2). In ecology, the **revolt arrows** in the panarchy model may be used to highlight these moments of realignment between different levels and rhythms (section 4.4.1).

| Terms used | Points for discussion | Fields of research |
|---|---|---|
| **Palimpsest (1950s on)** | Tablet or sheet of parchment, scraped to enable reuse ( Larousse 1898, p. 628; Gaffiot 1981, p. 1105). | History<br>Archeology<br>Geography |
| **Synchrony** | Axis of simultaneities, concerning the relations between coexistent things and excluding any notion of time (de Saussure 1986, c. V).<br>"When social intervention in the environment creates a new state, coherent with the social formation in question"[44] (Chouquer 2000, p. 126). | Linguistics<br>Archeogeography |
| **Revolt arrow** | Events which occur in intermediate and fast cycles can have an influence on slower levels (Holling *et al.* 2002, p. 76).<br>The small, fast level may transform the large, slow level (Holling 2006, pp. 20–21). | Ecology |
| **Taphochrony** | "Modality of rupture in time and in the accumulation space by the definitive burial of a spatial state, not followed by retrieval. This modality expresses a notion of irreversibility and is, in this sense, a source of historical knowledge with respect to radical bifurcations in systems"[45] (Chouquer 2003a, p. 302). | Archeogeography |

**Table 8.9.** *Ideas used in expressing and understanding the temporalities of reorganization in landscape forms and social-ecological systems (Robert 2019)*

44 *Lorsque l'intervention sociale sur le milieu crée un état nouveau, cohérent avec la formation sociale en question.*

45 *Modalité de rupture du temps et de l'espace d'accumulation par enfouissement définitif d'un état spatial qui ne provoque pas de reprise. Cette modalité exprime une irréversibilité et est, en ce sens, source de connaissance historique sur des bifurcations radicales des systèmes.*

The notion of **taphochrony** is used in archeogeography to describe transitions in which one plot or road network is replaced by another (section 3.6.1). However, the definitive nature of this disappearance can only be defined in relation to the moment of observation: forms still have the potential to be reactivated, following non-linear temporalities. We may thus add a temporal aspect to the definition of taphochrony: a spatial state is only considered to be definitively buried if no retrieval has occurred in the last *x* years, centuries, etc. (Table 8.9).

## 8.8. Conclusion: summary and directions for further investigation

To conclude this chapter, we shall highlight those notions studied in this book which we consider most promising with respect to the dynamics of forms in a landscape, viewed as a system. Table 8.10 shows the correspondences between different terms and notions used here, drawing attention to those elements that may be used to describe the dynamics of landscape forms in archeogeography, ecology and geography.

**Morphogenesis** may be used to describe the emergence of a plot or road network with the characteristics of a system. The *morpho* root recalls the importance of the morphological approach in archeogeographical analysis. Furthermore, the notion of morphogenesis is used in the field of complex systems.

With respect to the organization of plot or road networks as systems, the notions of **attractor** and **basin of attraction** are extremely helpful in the context of landscape form analysis, notably in defining connections with the notion of **morphogens**, used in archeogeography. Drawing on the work of C. Aschan-Leygonie, in the early 2000s, we applied these notions to a study of the long-term resilience of road networks. However, after a more detailed study of work by ecologists and geographers, this earlier approach seems overly allusive, and would merit further reflection and development over the coming years. The *Dynamique et Résilience des réseaux de circulation dans le Bassin parisien* (DYNAREP) program, supported by the ArcheoFab platform (UMR Arscan, Nanterre), gave us the opportunity to study the regional road network from ancient times to the present day, using both written and archeological sources. Our findings were systematically geolocated in a database of routes in the Parisian basin (Costa and Robert forthcoming). This database is intended to support the application of the notions of attractor and basin of attraction to the study of road networks in archeogeography.

| Field of reserch | Birth of a system | Organization | Articulation between persistence and change: resilience | Tipping/shifts | Reorganization | Inheritance, legacies, heritage | Potential for reuse |
|---|---|---|---|---|---|---|---|
| **Archeogeography** | Morphogenesis | Morphostasis, morphogens, morphological principles (isotopy, isoclination, iso-axiality) | (Diachrony), morphostasis, resilience, asynchrony of flow-ground footprint-shape, transformission, transmission | Synchrony, taphochrony, bifurcation | Transformission, morphogenesis | Hysterechrony | Uchrony |
| **Ecology** | ? | Attractor, basin of attraction, r phase, K phase | Resilience panarchy | Creative destruction, omega phase, release phase, regime shift, tipping point, revolt arrow | Alpha or reorganization phase | Hysteresis, path dependence, remember arrow | Remember arrow in panarchy model |

| **Field of reserch** | **Birth of a system** | **Organization** | **Articulation between persistence and change: resilience** | **Tipping/shifts** | **Reorganization** | **Inheritance, legacies, heritage** | **Potential for reuse** |
|---|---|---|---|---|---|---|---|
| **Geography** | Emergence, systemogenesis | Attractor | Resilience | Bifurcation, transition | Systemogenesis | Path dependence, heritage, spatial reproduction, spatial memory, inscribed memory | ? |
| **Landscape dynamics** | **Morphogenesis** | **Morphogen, spatial attractor, morphological principles (isotopy, isoclination, iso-axiality), basin of attraction** | **Resilience, panarchy** | **Transition** | **Morphogenesis** | **Hysteresis** | **Uchrony** |

**Table 8.10.** *Summary of correspondences between the main notions presented in this book. Our propositions for describing the dynamics of landscape forms, creating a common language for archeogeographers, ecologists and geographers, are shown in bold (Robert 2019)*

Given that morphogens generally correspond to geographical elements, the notion of **spatial attractors** offers a promising subject for further study. In ecology, this notion is used in understanding the movement of deep-sea species, and describes “environmental components which present the essential property of attracting individuals or groups of individuals, i.e. of generating orientation behaviors in their direction”[46]. Attractors generally take the form of clearly defined spatial sites (buoys, islands, etc.), localized using fixed geographic coordinates (Girard 2005, p. 39). In geography, Dendrinos (1996) used the notion of spatial attractors to characterize certain towns upon which multiple dimensions converge.

In the context of our work on the resilience of regional road networks, we also aim to develop a more precise definition of **transitions** and reorganizations in landscape forms. At what point do we consider that a network has disappeared for good, and that a new network has been reorganized? Are there any identifiable cases of alternative attractors? A further aim of our work is to develop a clearer understanding of the notions of hysteresis and path dependence as they relate to alternative attractors.

One of the defining characteristics of archeogeography lies in considering the reuse of previous forms following a hiatus. This approach is expressed through the notion of **uchrony**, another term which merits clarification, particularly in the way in which it relates to **hysteresis**, **path** dependence and the **remember arrows** in the **panarchy** model.

Table 8.10 provides a summary of key notions discussed here. The column, “Landscape dynamics”, highlights our preferred terms for use in archeogeography, intended to promote greater precision and to facilitate interactions with the ecology and geography communities.

46 *...des composantes de l’environnement dont la propriété essentielle est d’attirer des individus ou des groupes d’individus, c’est-à-dire de générer des comportements d’orientation dans leur direction.*

# Conclusion

The conceptual framework of resilience, which has its roots in the sciences of complexity, provides a powerful tool for analyzing landscapes. In this sense, a landscape is seen as a self-organized social-ecological system, combining persistence and change in its operation. Within this framework, archeogeographers, ecologists and geographers can work together using shared notions, concepts and terminology.

A landscape, perceived as a social-ecological system or anthroposystem, is an integrated whole; environments and societies are not considered as two distinct elements, but as parts of a system in constant interaction. The landscape itself is one of the most striking material manifestations of the integration of society and environment.

There is no stationary state of equilibrium, no "finished" reference state or norm to maintain or restore. Social-ecological systems are in a constant state of transformation, and change is inherent to the system. Perturbations come from both without and within, and the moment at which they occur largely determines their impact on the system. Instead of focusing on the scale of perturbations, we must consider the size of the basin of attraction, as it determines the resilience of a system. Understanding inherited processes that continue to have a degree of influence on the present, along with the trajectories of systems up to the present day, is crucial in studying resilience or transitions in landscapes.

The landscapes of the past are not definitively buried beneath those of the present, and they continue to play a role in orientation. All societies inherit spaces that impose certain constraints in terms of implantation; this notion has been described using a variety of terms (inheritance, legacy, spatial memory, inscribed memory, received space, etc.). In our view, the term **inherited spatial context** may be used here, echoing the notion of "context" described by F. Durand-Dastès[1]. Decisions made by an individual or group with respect to the creation of a new landscape element must take account of pre-existing structures in the host landscape, which are, in part, determined by a number of inherited morphological characteristics (orientations, sites, alignments, etc.) that ensure the resilience of plot and road network organizations. Inherited spatial context, like inscribed memory, consists of forms of both natural and social origin:

> Inscribed memory records the effects of the forces and times of nature, alongside those of past and bygone human actions, the concrete manifestations of which are quite as evident as those of natural processes: an urban network has just as much physical presence as a valley or vegetation pattern[2]. (Durand-Dastès 1990, p. 181)

This approach avoids the trap of natural determinism, as the inherited spatial context is also formed from social elements; furthermore, the trajectory of a system may follow any of a number of possible scenarios (uchrony).

The inherited spatial context imposes constraints on the present, but it may also be a source of resilience for present systems, introducing diversity and variety into a landscape. The creation of new road elements reusing inherited public ground footprints, for example, old paths, town boundary ditches, and so on, makes the circulation system more resilient on the macro level, while tying it even more closely to the micro level of towns, trading centers, etc. Agricultural systems may draw on existing irrigation or drainage networks, for example, in order to adapt (as in the case of the Comtat). The inherited spatial context thus has a non-negligible role to play in the resilience of a landscape as a social-ecological system.

Landscapes bear the marks of the ways in which past societies have adapted to change, and, as such, may provide valuable lessons in resilience. Thus, researchers working on the past have an active role to play in the debate surrounding societal

---

1 Durand-Dastès (2004) explains that microdecisions (such as an individual's decision to migrate) are oriented by "contexts", which depend on macrodecisions, such as the opening or closure of borders, settlement policies, economic disparities, and so on (p. 20).

2 *La mémoire inscrite enregistre les effets des forces et des temps de la nature, comme ceux des actions humaines de jadis et naguère, dont la traduction est aussi concrète, que celle des processus naturels : un réseau urbain a tout autant de présence physique qu'une vallée ou qu'une formation végétale.*

adaptation to environmental change, both by reconstructing the trajectories of social-ecological systems and by showing how past societies responded to similar challenges. Landscapes are illustrations of a shared experience, transmitted by the societies of the past, of the resilience of social-ecological systems characterized by a constant articulation of transformation and permanence.

# References


Abbé, J.-L. (2006). *À la conquête des étangs : l'aménagement de l'espace en Languedoc méditerranéen, XIIème–XVème siècles*. Presses universitaires du Mirail, Toulouse.

Adams, R.M. (1978). Strategies of maximization, stability, and resilience in Mesopotomian society, settlement, and agriculture. *Proceedings of the American Philosophical Society*, 122(5), 29–335.

Adger, W.N. (2000). Social and ecological resilience: Are they related? *Progress in Human Geography*, 24(3), 347–364.

Agache, R. (1978). *La Somme pré-romaine et romaine d'après les prospections aériennes à basse altitude.* Société des Antiquaires de Picardie, Amiens.

Altieri, M.A. (1995). *Agroecology: The Science of Sustainable Agriculture.* Westview Press, Boulder.

Alexander, D.E. (2013). Resilience and disaster risk reduction: an etymological journey. *NHESS*, 13(11), 2707–2716.

Ames, K.M. (1981). The evolution of social ranking on the northwest coast of North America. *American Antiquity*, 46(4), 789–805.

Amoros, C. and Petts, G.E. (eds) (1993). *Hydrosystèmes fluviaux.* Masson, Paris/Milan/Barcelona.

Archaeomedes (Durand-Dastès, F., Favory, F., Fiches, J.L., Mathian, H., Pumain, D., Raynaud, C., Sanders, L., Van der Leeuw, S.) (1998). *Des oppida aux métropoles : archéologues et géographes en vallée du Rhône.* Anthropos, Paris.

Archeologia. (1973). L'archéologie aérienne : vision fantastique du passé. *Archeologia*, 1.

Aschan-Leygonie, C. (1998). La Résilience d'un système spatial : l'exemple du Comtat. Une étude comparative de deux périodes de crises au XIXème et au XXème siècle. PhD Thesis, Université de Paris I, Paris.

Aschan-Leygonie, C. (2000). Vers une analyse de la résilience des systèmes spatiaux. *Espace géographique*, 29(1), 64–77.

Aston, M. and Rowley, T. (1974). *Landscape Archaeology: An Introduction to Fieldwork Techniques on Post-Roman Landscapes*. Newton Abbot/David & Charles, London/Vancouver.

Atelier d'architecture et d'urbanisme Treuttel Garcias Treuttel. (1996). Architecture des espaces publics modernes. De l'espace libre à l'espace public. Agency report.

Atlan, H. (1979). *Entre le cristal et la fumée. Essai sur l'organisation du vivant*. Le Seuil, Paris.

Aufrère, L. (1934). Les dunes du Sahara algérien (notes de morphologie dynamique). *Bulletin de l'Association de géographes français*, 11(83), 130–142.

Aufrère, L. (1935). Les systèmes agraires dans les îles Britanniques. *Annales de géographie*, 44(250), 385–409.

Auriac, F. (1983). *Système économique et espace : le vignoble languedocien*. Economica, Paris.

Bailey, G. (2007). Time perspectives, palimpsests and the archaeology of time. *Journal of Anthropological Archaeology*, 26(2), 198–223.

Barroca, B., DiNardo, M., Mboumoua, I. (2013). De la vulnérabilité à la résilience : mutation ou bouleversement ? *EchoGéo* [Online]. Available at: 10.4000/echogeo.13439 [Accessed 4 October 2018].

Baudelle, G. (1990). L'héritage spatial, agent d'inertie : l'exemple des bassins miniers. In *Géopoint 90 : Histoire, temps et espace*, Groupe Dupont (ed.). Université d'Avignon, Avignon.

Baudelle, G. and Regnauld, H. (2004). *Échelles et temporalités en géographie*. Sedes, Paris.

Beazeley, G.A. (1919). Air photography in archaeology. *The Geographical Journal*, 53(5), 330–335.

Beck, U. (2001). *La société du risque : sur la voie d'une autre modernité*. Aubier, Paris.

Beck, C., Luginbühl, Y., Muxart, T. (eds) (2006). *Temps et espaces des crises de l'environnement*. Quae, Versailles.

Bedon, R., Pinon, P., Chevallier, R. (1988). *Architecture et urbanisme en Gaule romaine. Tome 2. L'urbanisme en Gaule romaine*. Errance, Paris.

Bengtsson, J., Angelstam, P., Elmqvist, T., Emanuelsson, U., Folke, C., Ihse, M., Moberg, F., Nyström, M. (2003). Reserves, resilience and dynamic landscapes. *AMBIO: A Journal of the Human Environment*, 32(6), 389–396.

Beresford, M.W. and Hurst, J.G. (1971). *Deserted Medieval Villages: Studies*. Lutterworth Press, London.

Berger, J.-F. and Jung, C. (1996). Fonction, évolution et "taphonomie" des parcellaires en moyenne vallée du Rhône. Un exemple d'approche intégrée en archéo-morphologie et en géoarchéologie. In *Les Formes du paysage*, volume 2, Chouquer, G. (ed.). Errance, Paris.

Bergier, N. (1622). *Histoire des grands chemins de l'Empire romain*. C. Morel, Paris.

Berkes, F. (2009). Evolution of co-management: Role of knowledge generation, bridging organizations and social learning. *Journal of Environmental Management*, 90, 1692–1702.

Beroutchachvili, N. and Bertrand, G. (1978). Le géosystème ou "système territorial naturel". *Revue géographique des Pyrénées et du Sud-Ouest – Sud-Ouest Européen*, 49(2), 167–180.

Bertrand, G. (1991). La mémoire des terroirs. In *Pour une archéologie agraire : à la croisée des sciences de l'homme et de la nature*, Guilaine, J. (ed.). Armand Colin, Paris.

Bertrand, C. and Bertrand, G. (2002). *Une géographie traversière : l'environnement à travers territoires et temporalités*. Quae, Versailles.

Besse, J.-M. (2009). *Le goût du monde : exercices de paysage*. Actes Sud, Arles.

Bèzes, B., Verdier, N., Robert, S. (2014). La représentation des routes sur les cartes anciennes. In *Dynamique et résilience des réseaux routiers en région Île-de-France – Revue archéologique du Centre de la France*, Robert, S., Verdier, N. (eds). Librairie archéologique, Tours.

Bloch, M. (1934). Une étape de la seigneurie. L'exploitation de la réserve à Prüm, au IXème siècle. *Annales d'histoire économique et sociale*, 6(29), 450–489.

Bloch, M. (1966). *French Rural History: An Essay on its Basic Characteristics*. London, Routledge.

Bloch, M. (1988). *Les caractères originaux de l'histoire rurale française*. Armand Colin, Paris.

Boissinot, P. (2011). *L'archéologie comme discipline ?* Le Seuil, Paris.

Boissinot, P. and Brochier, J.-E. (1997). Pour une archéologie du champ. In *Les formes du paysage*, volume 2, Chouquer, G. (ed.). Errance, Paris.

Bourcart, L. (2015). Émergence et usages du concept de résilience dans les mondes académique et institutionnel. PhD Thesis, Université Grenoble Alpes, Grenoble.

Bourgine, P. and Lesne, A. (2006). *Morphogenèse : l'origine des formes*. Belin, Paris.

Braje, T.J., Erlandson, J.M., Rick, T.C., Dayton, P.K., Hatch, M.B.A. (2009). Fishing from past to present: Continuity and resilience of Red Abalone Fisheries on the Channel Islands, California. *Ecological Applications*, 19(4), 906–919.

Brand, F. and Jax, K. (2007). Focusing the meaning(s) of resilience: Resilience as a descriptive concept and a boundary object. *Ecology and Society*, 12(1) [Online]. Available at: https://www.ecologyandsociety.org/vol12/iss1/art23/.

Braudel, F. (1958). La longue durée. *Annales d'histoire économique et sociale*, 4, 725–753.

Bravard, J.-P. and Magny, M. (2002). *Les fleuves ont une histoire : paléo-environnement des rivières et des lacs français depuis 15 000 ans*. Errance, Paris.

Brigand, R. (2010). Centuriations romaines et dynamique des parcellaires. Une approche diachronique des formes rurales et urbaines de la plaine centrale de Venise (Italie). PhD Thesis, Université de Franche-Comté, Università degli Studi di Padova, Besançon.

Briggs, J.M., Spielmann, K.A., Schaafsma, H., Kintigh, K.W., Kruse, M., Morehouse, K., Schollmeyer, K. (2006). Why ecology needs archaeologists and archaeology needs ecologists. *Frontiers in Ecology and the Environment*, 4(4), 180–188.

Brinckmann, A.E. (1908). *Platz und Monument: Untersuchungen zur Geschichte und Ästhetik der Stadtbaukunst in neuerer Zeit.* Wasmuthedit, Berlin.

Brookfield, H.C. (1984). Intensification revisited. *Pacific Viewpoint*, 25(1), 15–44.

Brown, K. (2014). Global environmental change I: A social turn for resilience? *Progress in Human Geography*, 38(1), 107–117.

Bruant, J., Cotté, O., Pissot, V. (2014). La voie antique Orléans-Paris à Massy-Palaiseau (Essonne). In *Dynamique et résilience des réseaux routiers – 52ème supplément à la Revue archéologique du Centre de la France*, Robert, S., Verdier, N. (eds). FERACF, Tours.

Brunet, R. (1974). Analyse des paysages et sémiologie. Éléments pour un débat. *Espace géographique*, 3(2), 120–126.

Brunet, R. (2001). *Le déchiffrement du monde.* Belin, Paris.

Brunet, R., Ferras, R., Théry, H. (eds). (1993). *Les mots de la géographie : dictionnaire critique.* Reclus, Montpellier.

Brunhes, J. (1925). *La géographie humaine.* Félix Alcan, Paris.

Brunhes, J. and Vallaux, C. (1921). *La géographie de l'histoire : géographie de la paix et de la guerre sur terre et sur mer.* Félix Alcan, Paris.

Brunhes-Delamarre, M. and Deffontaines, P. (1955). *Atlas aérien de la France. 5 tomes.* Gallimard, Paris.

Bur, M. (ed.). (1993). *Les peuplements castraux dans les pays de l'Entre-Deux : Alsace, Bourgogne, Champagne, Franche-Comté, Lorraine, Luxembourg, Rhénanie-Palatinat, Sarre. Actes du colloque de Nancy, 1–3 octobre 1992.* Presses universitaires de Nancy, Nancy.

Burel, F. and Baudry, J. (1999). *Écologie du paysage. Concepts, méthodes et applications.* Tec & Doc, Paris.

Bürgi, M., Östlund, L., Mladenoff, D.J. (2017). Legacy effects of human land use: Ecosystems as time-lagged systems. *Ecosystems*, 20(1), 94–103.

Burnouf, J. and Gillot, L. (2016). L'archéogéographie : pourquoi ? Comment ? Entretien avec Hélène Noizet et Sandrine Robert. *EchoGéo* [Online]. Available at: 10.4000/echogeo.14633.

Burnouf, J., Carcaud, N., Garcin, M. (2009). Nouvelle histoire pour la Loire entre nature et société. *Geosciences*, 2009, 72–79.

Butzer, K.W. (1980). Context in archaeology: An alternative perspective. *Journal of Field Archaeology*, 7(4), 417–422.

Caillemer, A. and Chevallier, R. (1958). Mise au point à propos de l' "Atlas des centuriations romaines de Tunisie" : quelques questions de méthode. *Annales. Histoire, Sciences sociales*, 13(2), 304–307.

Campbell, S. and Butler, V. (2010). Archaeological evidence for resilience of Pacific Northwest salmon populations and the socioecological system over the last ~7,500 years. *Ecology and Society*, 15(1) [Online]. Available at: https://www.ecologyandsociety.org/vol15/iss1/art17/ [Accessed 15 September 2017].

Cange, D. (1678). *Glossarium Mediae et Infimae Latinitatis*, Volume 2, Basel.

Carpenter, S. (2011). What is resilience? Stockholm resilience centre [Online]. Available at: http://www.stockholmresilience.org/21/research/research-videos/12-1-2011-what-is-resilience-.html [Accessed 23 November 2015].

Carpenter, S., Walker, B., Anderies, J.M., Abel, N. (2001). From metaphor to measurement: Resilience of what to what? *Ecosystems*, 4(8), 765–781.

Cavanna, E. (2016). Spatialisation des élites rurales médiévales et modernes dans le Bassin parisien. Pour une approche archéogéographique des pratiques sociales de distinction. PhD Thesis, Université Paris I Panthéon Sorbonne, Paris [Online]. Available at: https://tel.archives-ouvertes.fr/tel-01381024/document [Accessed 1 October 2018].

Cerdà, I. (1979). *La théorie générale de l'urbanisation. Présentée et adaptée par Lopez de Aberasturi.* Le Seuil, Paris.

Cerdà, I. (2018). *General Theory of Urbanization 1867*, Volume 1. Translated by Vicente G. Institute for Advanced Architecture of Catalonia, Barcelona.

Chandler, D. and Coaffee, J. (eds). (2017). *The Routledge Handbook of International Resilience.* Taylor & Francis, London.

Chapelon, L. (2004). Réseau, Hypergeo [Online]. Available at: http://www.hypergeo.eu/spip.php?article80 [Accessed 24 October 2018].

Charpy, G. (1901). Note sur l'essai des métaux à la flexion par choc de barreaux entaillés. *Mémoires de la société des ingénieurs civils*, 76, 848–877.

Charvet, J.-P. (1985). *Les Greniers du monde.* Economica, Paris.

Chesneau, I. and Roncayolo, M. (2011). *L'abécédaire de Marcel Roncayolo. Entretiens.* Infolio, Paris.

Chevallier, R. (1956). *Atlas des centuriations romaines de Tunisie.* IGN, Paris.

Chevallier, R. (ed.). (1964). *Colloque international d'archéologie aérienne.* SVEPEN, Paris.

Chevallier, R. (1971). *La photographie aérienne.* Armand Colin, Paris.

Chevallier, R. (1976). Le paysage palimpseste de l'histoire : pour une archéologie du paysage. *Mélanges de la Casa de Velázquez*, 12(1), 503–510.

Chevallier, R. (2000). *Lecture du temps dans l'espace : topograhie archéologique et historique.* Picard, Paris.

Chevillot-Miot, E. (2017). La résilience des territoires littoraux face au risque de submersion marine. Application sur les territoires de la Charente-Maritime et de la Somme. PhD Thesis, Université de Nantes, Nantes [Online]. Available at: https://tel.archives-ouvertes.fr/tel-01659505/document [Accessed 18 June 2018].

Choay, F. (1965). *L'urbanisme, utopies et réalistés. Une anthologie.* Le Seuil, Paris.

Choay, F. and Merlin, P. (eds). (1988). *Dictionnaire de l'urbanisme et de l'aménagement.* Presses universitaires de France, Paris.

Chombart de Lauwe, P.H. (ed.). (1948). *La découverte aérienne du monde.* Horizons de France, Paris.

Chouquer, G. (1983). Dégradation et fossilisation des cadastres centuriés, étude morphologique. In *Cadastres et espace rural. Actes de la table ronde de Besançon (mai 1980)*, Clavel-Lévêque, M. (ed.). CNRS, Paris.

Chouquer, G. (1990). Cours d'archéomorphologie en trois tomes. Report, CNRS, Tours.

Chouquer, G. (1994). Actualité des recherches sur les parcellaires ruraux et urbains. *Les Nouvelles de l'Archéologie*, 56, 25–32.

Chouquer, G. (ed.). (1996a). *Les formes du paysage. Tome 1 : Étude sur les parcellaires.* Errance, Paris.

Chouquer, G. (ed.). (1996b). *Les formes du paysage. Tome 2 : Archéologie des parcellaires.* Errance, Paris.

Chouquer, G. (1996c). Parcellaires et longue durée. Points de repère historiques et problèmes d'interprétation. In *Les formes du paysage*, volume 2, Chouquer, G. (ed.). Errance, Paris.

Chouquer, G. (1997a). La place de l'analyse des systèmes spatiaux dans l'étude des paysages du passé. In *Les formes du paysage*, volume 3, Chouquer, G. (ed.). Errance, Paris.

Chouquer, G. (ed.). (1997b). *Les formes du paysage. Tome 3 : L'analyse des systèmes spatiaux.* Errance, Paris.

Chouquer, G. (2000). *L'étude des paysages : essais sur leurs formes et leur histoire.* Errance, Paris.

Chouquer, G. (2003a). Crise et recomposition des objets : les enjeux de l'archéogéographie. *Études rurales*, 167–168, 13–31.

Chouquer, G. (ed.). (2003b). Objets en crise. Objets recomposés. Enjeux et contours de l'archéogéographie. *Études rurales*, 2003, 13–307.

Chouquer, G. (2005). L'émergence de la planimétrie agraire à l'âge du fer. *Études rurales*, 175–176, 29–81.

Chouquer, G. (2007a). *Quels scénarios pour l'histoire du paysage ? Orientation de recherche pour l'archéogéographie, essai.* Universités de Coimbra et Porto, Coimbra/Porto.

Chouquer, G. (2007b). Transmissions et transformations dans les formes parcellaires en France. Esquisse d'un schéma général d'interprétation. In *La mémoire des forêts. Actes du colloque Forêt, archéologie et environnement*, Dupouey, J.-L., Dambrine, É., Dardignac, C., Georges-Leroy, M. (eds). Office national des forêts, Institut national de la recherche agronomique et Direction régionale des affaires culturelles de Lorraine, Metz.

Chouquer, G. (2008a). Les transformations récentes de la centuriation. *Annales. Histoire, Sciences sociales*, 63(4), 847–874.

Chouquer, G. (2008b). *Traité d'archéogéographie.* Errance, Paris.

Chouquer, G. (2009). Ce que le temps fait aux formes planimétriques. *Géocarrefour*, 84(4), 217–226.

Chouquer, G. (2011). Une discipline sans mandat. *Études rurales*, 188(2), 179–187.

Chouquer, G. and Watteaux, M. (2013). *L'archéologie des disciplines géohistoriques.* Errance, Paris.

Clark, M.J. (1998). Flood insurance and coastal resilience: A UK-focused discussion. *The Geographical Journal*, 164(3), 333–334.

Cohen, J.-L. (1993). L' "architecture urbaine" selon P. Lavedan. *Ville et Architecture*, 32–33, 157–167.

Comfort, L.K., Boin, A., Demchak, C.C. (eds). (2010). *Designing Resilience. Preparing for Extreme Events.* University of Pittsburgh Press, Pittsburgh.

Conolly, J. and Lane, P.J. (eds). (2018). Vulnerability, risk and resilience: An introduction. *World Archaeology*, 50(4), 547–553.

Corboz, A. (1983). Le territoire comme palimpseste. *Diogène*, 121, 14–35.

Costa, L. and Robert, S. (forthcoming). Constitution d'un référentiel sur les voies antiques dans le programme DYNAmique et REsilience des réseaux de circulation dans le Bassin Parisien (DYNAREP). *NUMEAR, Archéologies numériques*, ISTE Openscience.

Costa, L., Pariente, A., Robert, S. (2014). SIGArgos, de Vollgraff au SIG, ou l'espace argien revisité. In *Recherches franco-helléniques IV, École française d'Athènes. Sur les pas de W. Vollgraff. Cent ans d'activités archéologiques à Argos. Actes du colloque international organisé par Ive EPKA et l'École française d'Athènes 25–28 sept. 2003*, Mulliez, D. (ed.). École française d'Athènes, Athens.

Cotelle, P. (2014). Une géographie de l'insécurité urbaine post catastrophe : le cas de la Nouvelle-Orléans et du cyclone Katrina (États-Unis, 2000–2010). PhD Thesis, Université Paul Valéry, Montpellier.

Crawford, O.G.S. (1923). Air survey and archaeology. *Geographical Journal*, 1923, 342–360.

Crawford, O.G.S. (1929). *Air-photography for Archaeologists.* His Majesty's Stationery Office, London.

Crawford, O.G.S. (1953). *Archaeology in the Field.* Phoenix House, London.

Crawford, O.G.S. and Keiller, A. (1928). *Wessex from the Air*. Clarendon Press, Oxford [Online]. Available at: http://www.dli.ernet.in/handle/2015/282493.

Cretney, R. (2014). Resilience for whom? Emerging critical geographies of socio-ecological resilience. *Geography Compass*, 8(9), 627–640.

Crumley, C.L. (2012). A heterarchy of knowledge: Tools for the study of landscape histories and futures. In *Resilience and the Cultural Landscape: Understanding and Managing Change in Human-shaped Environments*, Plieninger, T., Bieling, C. (eds). Cambridge University Press, Cambridge.

Cyrulnik, B. (2002). *Un merveilleux malheur*. Odile Jacob, Paris.

Cyrulnik, B. and Jorland, G. (eds). (2012). *Résilience, connaissances de base*. Odile Jacob, Paris.

D'Amico, J.C. (2009). *Le mythe impérial et l'allégorie de Rome. Entre Saint-Empire, Papauté et Commune*. Presses universitaires de Caen, Caen.

D'Urban, A.J.F. (1837). *Table générale, alphabétique et analytique, des matières contenues dans les quinze premiers tomes formant seize volumes de l'Histoire de Hainaut de Jacques de Guyse*. Chez l'auteur, Paris.

Darby, H.C. and Williams, M. (2002). *The Relations of History and Geography: Studies in England, France and the United States*. University of Exeter Press, Exeter.

Darin, M. (1998). The study of urban form in France. *Urban Morphology*, 2(2), 63–76.

Dauphiné, A. (1998). Espace terrestre et espace géographique. In *Les concepts de la géographie humaine*, Bailly, A. (ed.). Armand Colin, Paris.

Dauphiné, A. and Provitolo, D. (2007). La résilience : un concept pour la gestion des risques. *Annales de géographie*, 654(2), 115–125.

Demangeot, J. (1943). Notes sur la haute vallée de l'Ubaye. Structure, altitudes moyennes, tracé du réseau hydrographique. *Revue de géographie alpine*, 31(4), 535–574.

Dendrinos, D.S. (1996). Cities as spatial chaotic attractors. In *Chaos Theory in the Social Sciences: Foundations and Applications*, Kiel, E. and Elliott, E. (eds). Michigan University Press, Ann Arbor.

Denevan, W.M. (1995). Prehistoric agricultural methods as models for sustainability . In *Advances in Plant Pathology*, Andrews, J.H., Tommerup, I.C. (eds). Academic Press, London, 21–43 [Online]. Available at: http://www.sciencedirect.com/science/article/pii/S0736453906800048 [Accessed 9 January 2017].

Derruau, M. (1996). *Les formes du relief terrestre. Notions de géomorphologie*. Armand Colin, Paris.

Desachy, B. (2008). De la formalisation du traitement des données stratigraphiques en archéologie de terrain. PhD Thesis, Université Panthéon-Sorbonne, Paris.

Devals, C. (2003). Marines-Santeuil – Les Carreaux. Document final de synthèse de fouille préventive de mars à août 2001. Report, DRAC-SRA-Île-de-France, INRAP, CG-Val-d'Oise, Saint-Denis, Pantin.

Diamond, J.M. (2005). *Collapse: How Societies Choose to Fail Or Succeed*. Penguin Books, London.

Dion, R. (1949). La géographie rétrospective, leçon inaugurale au collège de France. *Cahiers internationaux de sociologie*, VI, 3–27.

Djament, G. (2005). La reproduction de la centralité romaine de la "Ville éternelle" à la capitale de l'Italie : essai de géohistoire urbaine. PhD Thesis, Université Paris VII – Diderot, Paris.

Djament-Tran, G. (2011). *Rome éternelle : les métamorphoses de la capitale*. Belin, Paris.

Djament-Tran, G. (2015). Resilience: A question of scale. In *Resilience Imperative: Uncertainty, Risks and Disasters*, Reghezza-Zitt, M. and Rufat, S. (eds). ISTE Press, London, and Elsevier, Oxford.

Djament-Tran, G. and Reghezza-Zitt, M. (eds). (2012). *Résiliences urbaines : les villes face aux catastrophes*. Le Manuscrit, Paris.

Ducrot, O. and Schaeffer, J.-M. (1995). *Nouveau dictionnaire encyclopédique des sciences du langage*. Points, Paris.

Ducrot, O. and Todorov, T. (1972). *Dictionnaire encyclopédique des sciences du langage*. Le Seuil, Paris.

Durand-Dastès, F. (1984). La question "Où ?" et l'outillage géographique. *Espaces temps, les cahiers*, 26(1), 8–21.

Durand-Dastès, F. (1986). L'interaction espaces/sociétés, la géographie sociale et la "théorie du renversement". *L'Espace géographique*, 15(2), 114–117.

Durand-Dastès, F. (1990). L'espace et les mémoires du monde. *Revue géographique des Pyrénées et du Sud-Ouest*, 61(2), 181–182.

Durand-Dastès, F. (1991). La géographie et les mémoires de Gaïa. *Géopoint 1990*, 141–146.

Durand-Dastès, F. (1992). Prise de position, Systèmes et modèles, La mémoire de Gaïa, Larmes et lecture de cartes, Le particulier et le général et autres textes, textes rassemblés par C. Grataloup et L . Sanders, CD Rom, Systèmes et modèles.

Durand-Dastès, F. (1999). Jamais deux fois... ou : de quelques précautions à prendre avec le temps. *Travaux de l'Institut de géographie de Reims*, 101–104, 5–23.

Durand-Dastès, F. (2001). Le temps, la géographie et ses modèles. *Bulletin de la société géographique de Liège*, 40(1), 5–13.

Durand-Dastès, F. (2004). Temporalités et échelles : définitions et premières articulations. Chapitre premier. *Cours de géographie*. Centre national d'enseignement à distance, Vanves, 70.

Durand-Dastès, F. (2010). Ordres de grandeur et systèmes. Conference, *Géopoint 2010. Les échelles pour les géographes et les autres*, 3–4 June.

Durand-Dastès, F. and Sanders, L. (2004). Système spatial. *Hypergéo* [Online]. Available at: http://www.hypergeo.eu/spip.php?article85 [Accessed 3 April 2018].

Elissalde, B. (2000). Géographie, temps et changement spatial. *Espace géographique*, 29(3), 224–236.

Encyclopedia Universalis (2001). *Dictionnaire de la physique mécanique et thermo-dynamique*. Encyclopaedia universalis. Albin Michel, Paris.

Fabiani, J.-L. (2016). *Pierre Bourdieu. Un structuralisme héroïque*. Le Seuil, Paris.

Faulseit, R.K. (ed.). (2016). *Beyond Collapse: Archaeological Perspectives on Resilience, Revitalization, and Transformation in Complex Societies*. Southern Illinois University Press, Carbondale.

Favory, F. (1997). Retour critique sur les centuriations du Languedoc oriental, leur existence et leur datation. In *Les Formes du paysage*, volume 3, Chouquer, G. (ed.). Errance, Paris.

Favory, F. (2012). Les parcellaires antiques de Gaule médiane et septentrionale. In *Des hommes aux champs. Pour une archéologie des espaces ruraux du Néolithique au Moyen Âge*, Charpentier, V., Marcigny, C. (eds). Presses uni-versitaires de Rennes, Rennes [Online]. Available at: https://hal.campus-aar.fr/medihal-01893971/document [Accessed 7 January 2019].

Favory, F., Nuninger, L., Bertoncello, F., Fovet, E., Gauthier, E., Poirier, N., George-Leroy, M., Ostir, K., Saligny, L., Trément, F., Weller, O. (2008). ArchacDyn. Dynamique spatiale du peuplement et ressources naturelles : vers une analyse intégrée dans le long terme, de la Préhistoire au Moyen Âge. ACI Espaces et territoires (contrat ET28, 2005–2007). Report, CNRS, Besançon [Online]. Available at: https://halshs.archives-ouvertes.fr/halshs-00483141 [Accessed 24 June 2019].

Favory, F., Nuninger, L., Sanders, L. (2012). Intégration de concepts de géographie et d'archéologie spatiale pour l'étude des systèmes de peuplement. *L'Espace géographique*, 41(4), 295–309.

Feichtinger, G., Forst, C.V., Piccardi, C. (1996). A nonlinear dynamical model for the dynastic cycle. *Chaos, Solitons & Fractals*, 7(2), 257–269.

Ferdière, A. (1983). Recherche des parcellaires fossiles et cadastres antiques : l'exemple de la Gaule du Nord. In *Cadastres et espace rural. Actes de la table ronde de Besançon (mai 1980)*, Clavel-Lévêque, M. (ed.). CNRS, Paris.

Fisher, C.T. and Feinman, G.M. (2005). Introduction to landscapes over time: Resilience, degradation, and contemporary lessons. *American Anthropologist*, 107(1), 62–69.

Fisher, C.T., Brett Hill, J., Feinman, G.M. (2009). *Exploring Solutions Past: The Role of Archaeology in Shaping Conservation Strategies of the Future. The Archaeology of Environmental Change: Socionatural Legacies of Degradation and Resilience*. University of Arizona Press, Tucson.

Folke, C. (1997). Ecosystem appropriation by cities. *Ambio*, 26, 167–172.

Folke, C. (2011). Learning and adapting: Central features of resilience. In *Resilience in Social-Ecological Systems: The Role of Learning and Education*, Krasny, M.E., Lundholm, C., Plummer, R. (eds). Routledge, London [Online]. Available at: http://urn.kb.se/resolve?urn=urn:nbn:se:su:diva-66718 [Accessed 26 April 2018].

Folke, C. (2016). Resilience. In *Oxford Research Encyclopedia of Environmental Science* [Online]. Available at: https://oxfordre.com/environmentalscience/environmentalscience/view/10.1093/acrefore/9780199389414.001.0001/acrefore-9780199389414-e-8.

Folke, C. and Berkes, F. (eds). (1998). *Linking Social and Ecological Systems: Management Practices and Social Mechanisms for Building Resilience*. Cambridge University Press, Cambridge.

Folke, C., Carpenter, S., Walker, B., Scheffer, M., Chapin, T., Rockström, J. (2010). Resilience thinking: Integrating resilience, adaptability and transformability. *Ecology and Society*, 15(4) [Online]. Available at: https://www.ecologyandsociety.org/vol15/iss4/art20/ [Accessed 15 December 2017].

Foster, D., Swanson, F., Aber, J., Burke, I., Brokaw, N., Tilman, D., Knapp, A. (2003). The importance of land-use legacies to ecology and conservation. *BioScience*, 53(1), 77–88.

Found, W. and Berbès-Blàzquez, M. (2012). The sugar-cane landscape of the Caribbean islands: Resilience, adaptation and transformation of the plantation social-ecological system. In *Resilience and the Cultural Landscape: Understanding and Managing Change in Human-shaped Environments*, Plieninger, T., Bieling, C. (eds). Cambridge University Press, Cambridge.

Fovet, É. (2010). Dynamiques socio-environnementales durant l'Antiquité : approche micro-régionale du peuplement en Languedoc oriental. PhD Thesis, Université de Franche-Comté, Besançon.

François, C. (2011). *International Encyclopedia of Systems and Cybernetics*. Walter de Gruyter, Berlin.

Fusco, G. (2018). Ville, complexité, incertitude. Enjeux de connaissance pour le géographe et l'urbaniste. PhD Thesis, HDR, Université de Nice Sophia Antipolis, Nice [Online]. Available at: https://hal.archives-ouvertes.fr/tel-01968002 [Accessed 25 June 2019].

Gaffiot, F. (1981). *Dictionnaire illustré Latin-Français*. Hachette, Paris.

Gaillard, J.C. (2010). Vulnerability, capacity and resilience: Perspectives for climate and development policy. *Journal of International Development*, 22(2), 218–232.

Gandini, C. (2006). Des campagnes gauloises aux campagnes de l'Antiquité tardive : la dynamique de l'habitat rural dans la cité des Bituriges Cubi (IIe s. av. J.-C.-VIIe s. ap. J.-C.). PhD Thesis, Université Panthéon-Sorbonne – Paris I, Paris [Online]. Available at: http://tel.archives-ouvertes.fr/tel [Accessed 25 June 2019].

Garcia, P.-O. (2015). Sous l'adaptation, l'immunité. Étude sur le discours de l'adaptation au changement climatique. PhD Thesis, Université de Grenoble, Grenoble [Online]. Available at: https://tel.archives-ouvertes.fr/tel-01245377/file/These-GarciaPO-2015.pdf [Accessed 19 June 2019].

Garmy, P. (2009). Villes, réseaux et systèmes de villes : contribution de l'archéologie, volume 2. HDR, Université de Paris I, Paris [Online]. Available at: https://tel.archives-ouvertes.fr/tel-00459332 [Accessed 24 June 2019].

Gauthiez, B. (1999). Approche morphologique des plans de villes, archéologie et sources écrites. *Revue de Picardie*, 16, 17–27.

Georges, P. (1974). *L'ère des techniques : constructions ou destructions ?* Presses universitaires de France, Paris.

Georges, P. and Verger, F. (eds). (2006). *Dictionnaire de la géographie*. Presses universitaires de France, Paris.

Giligny, F. and Tsirtsoni, Z. (2016). Environnement et sociétés. *Les Nouvelles de l'archéologie*, 3–4.

Girard, C. (2005). Étude du comportement d'orientation d'espèces pélagiques tropicales vis-à-vis d'attracteurs. PhD Thesis, Université de la Réunion, Saint-Denis.

Godefroy, F. (1881). *Dictionnaire de l'ancienne langue française et de tous ses dialectes du IX$^e$ au XV$^e$ siècle : composé d'après le dépouillement de tous les plus importants documents manuscrits ou imprimés qui se trouvent dans les grandes bibliothèques de la France et de l'Europe municipales ; complement, inaccoutumé-zoophyte.* F. Vieweg (E. Bouillon), Paris.

González Villaescusa, R. (1996). Centuriations, alquerias et pueblas : éléments pour la compréhension du paysage valencien. In *Les formes du paysage*, volume 2, Chouquer, G. (ed.). Errance, Paris.

Grataloup, C. (1996). *Lieux d'histoire : essai de géohistoire systématique.* GIP Reclus, Montpellier.

Grenier, A. (1934). *Manuel d'archéologie gallo-romaine. Deuxième partie. L'archéologie du sol. Les routes.* Picard, Paris.

Grosso, E., Noizet, H., Robert, S., Chareille, P. (2013). The ALPAGE historical GIS: A new tool allowing a new look at medieval Paris. In *Understanding Landscapes: From Land Discovery to Spatial Organization, Proceedings of the Sessions C19 and C22 of the XVI World Congress (Florianópolis, 4–10 September 2011)*, Djindjian, F., Robert S. (eds). British Archeological Report, Oxford.

Guillot, J.-F. (2014). La ville et son élan vital. Marcel Poëte et Robert Auzelle lecteurs de Bergson. *Les Cahiers de la recherche architecturale et urbaine*, 123–130.

Gumerman, G.J. and Gell-Mann, M. (1994). *Understanding Complexity in the Prehistoric Southwest.* Addison-Wesley, New York [Online]. Available at: http://tuvalu.santafe.edu/~mgm/Site/Publications_files/MGM%20114.pdf [Accessed 26 June 2018].

Gunderson, L.H. and Holling, C.S. (eds) (2002). *Panarchy: Understanding Transformations in Human and Natural SYSTEMS.* Island Press, Washington.

Gunderson, L.H., Holling, C.S., Pritchard, L., Peterson, G.D. (1997). Resilience in ecosystems, institutions and societies. Discussion Paper 95, Stockholm.

Gunderson, L.H., Holling, C.S., Peterson, G.D. (2002). Surprises and sustainability: Cycles of renewal in the everglades. In *Panarchy: Understanding Transformations in Human and Ecological Systems*, Gunderson, L.H., Holling, C.S. (eds). Island Press, Washington.

Gunderson, L.H., Kinzig, A.P., Quinlan, A., Walker, J.H. (2010). Assessing resilience in social-ecological systems [Online]. Avaialble at: https://www.resalliance.org/files/ResilienceAssessmentV2_2.pdf [Accessed 26 April 2018].

Gunderson, L.H., Cosens, B.A., Chaffin, B.C., Arnold, C.A.(T.), Fremier, A.K., Garmestani, A.S., Craig, R.K., Gosnell, H., Birge, H.E., Allen, C.R., Benson, M.H., Morrison, R.R., Stone, M.C., Hamm, J.A., Nemec, K., Schlager, E., Llewellyn, D. (2017). Regime shifts and panarchies in regional scale social-ecological water systems. *Ecology and Society*, 22(1) [Online]. Available at: https://www.ecologyandsociety.org/vol22/iss1/art31/ [Accessed 26 April 2018].

Gunnell, Y. (2009). *Écologie et société : repères pour comprendre les questions d'environnement.* Armand Colin, Paris.

Guttmann-Bond, E.B. and Kluiving, S.J. (2015). *Landscape Archaeology between Art and Science: From a Multi- to an Interdisciplinary Approach.* Amsterdam University Press, Amsterdam.

Guy, M. and Passelac, M. (1991). Prospection aérienne et télédétection des structures de parcellaires. In *Pour une archéologie agraire*, Guilaine, J. (ed.). Armand Colin, Paris.

Haggett, P. (1965). *Locational Analysis in Human Geography.* E. Arnold, London.

Håkansson, N.T. and Widgren, M. (eds). (2014). *Landesque Capital: The Historical Ecology of Enduring Landscape Modifications.* Left Coast Press, Walnut Creek.

Halvorsen, R. and Layton, D.F. (2015). *Handbook on the Economics of Natural Resources.* Edward Elgar Publishing, Cheltenham/Camberley.

Hegmon, M., Peeples, M.A., Kinzig, A.P., Kulow, S.R., Meegan, C.M., Nelson, M.C. (2008). Social transformation and its human costs in the prehispanic U.S. Southwest. *American Anthropologist New Series*, 110(3), 313–324.

Helmer, W., Vellinga, P., Litjens, G., Goosen, H., Ruijgrok, E. (1996). Growing with the sea – Creating a resilient coastline [Online]. Available at: http://www.wwf.de/fileadmin/fm-wwf/Publikationen-PDF/WWF-NL_1996_Growing_with_the_Sea.pdf [Accessed 26 April 2018].

Hermenault, L. (2017). La ville en mouvements : circulations, échanges commerciaux et matérialité de la ville : pour une articulation systémique des facteurs d'évolution du tissu urbain parisien entre le XVe et le XIXe siècle. PhD thesis, Université Panthéon-Sorbonne – Paris I, Paris [Online]. Available at: https://tel.archives-ouvertes.fr/tel-01656824/document [Accessed 13 December 2018].

Hernandez, J. (2010). ReNew Orleans ? Résilience urbaine, mobilisation civique et création d'un "capital de reconstruction" à la Nouvelle-Orléans après Katrina. PhD Thesis, Université de Paris X, Nanterre [Online]. Available at: http://www.theses.fr/2010PA1 00224 [Accessed 12 June 2018].

Holling, C.S. (1965). The functional response of predators to prey density and its role in mimicry and population regulation. *Mem. Ent. Soc. Can.*, 45, 1–60.

Holling, C.S. (1973). Resilience and stability of ecological systems. *Annual Review of Ecology and Systematics*, 4, 1–23.

Holling, C.S. (ed.). (1978). *Adaptative Environmental Assessment and Management.* John Wiley and Sons, Chichester/New York/Brisbane/Toronto.

Holling, C.S. (1986). The resilience of terrestrial ecosystems; local surprise and global change. In *Sustainable Development of the Biosphere*, Clark, W.C., Munn, R.E. (eds). Cambridge University Press, Cambridge.

Holling, C.S. (1992). Cross-scale morphology, geometry and dynamics of ecosystems. *Ecological Monographs*, 62(4), 447–502.

Holling, C.S. (1996). Engineering resilience versus ecological resilience. In *Engineering within Ecological Constraints*, Schulze, P. (ed.). The National Academies Press, Washington, 31–44 [Online]. Available at: https://www.nap.edu/read/4919/chapter/4 [Accessed 3 April 2017].

Holling, C.S. (2006). A journey of discovery [Online]. Available at: http://www.resalliance.org/ files/Buzz_Holling_Memoir_2006_a_journey of discovery buzz holling.pdf [Accessed 26 December 2018].

Holling, C.S. and Gunderson, L.H. (2002). Resilience and adaptive cycles. In *Panarchy: Understanding Transformations in Human and Ecological Systems*, Gunderson, L.H., Holling, C.S. (eds). Island Press, Washington.

Holtorf, C. (2018). Embracing change how cultural resilience is increased through cultural heritage. *World Archaeology*, 50(4), 639–650.

Hopkins, R. (2008). *The Transition Handbook. From Oil Dependency to Local Resilience.* Green Books, Cambridge.

Janssen, M.A. and Ostrom, E. (2006). Resilience, vulnerability, and adaptation: A cross-cutting theme of the International Human Dimensions Programme on Global Environmental Change. *Global Environmental Change*, 16(3), 237–239.

Jobic, F. (2001). Courcelles-sur-Viosne (Val-d'Oise), "La Chaussée Jules César". Document final de synthèse de fouille préventive du 20/01/2000 au 31/03/2000. Report, SRA Ile-de-France, AFAN, SDAVO, Saint-Denis.

Johnson, M. (2008). *Ideas of Landscape.* John Wiley & Sons, Hoboken.

Jones, C.W. (ed.). (1977). *Bède le Vénérable, De temporum ratione liber.* Brepols, Turnhout.

Jorré, G. (1933). Gel du sol et déflation par le froid en montagne. Le problème des "Terrasses Goletz" sibériennes (d'après quelques travaux russes récents). *Revue de géographie alpine*, 21(2), 347–371.

Juillard, E. (1961). La genèse des paysages agraires. *Annales. Histoire, Sciences sociales*, 16(5), 993–1003.

Jullian, C. (1964). *Histoire de la Gaule. Tome 5 – La civilisation gallo-romaine. État matériel*. Hachette, Paris.

Jung, C. (1999). Morphogenèse, fonctions et évolution de la centuriation B d'Orange et essai de restitution diachronique des paléopaysages du Tricastin (Drôme-Vaucluse). PhD Thesis, Université François Rabelais, Tours.

Kinzig, A. (2012). Towards a deeper understanding of the social resilience: The contributions of cultural landscapes. In *Resilience and the Cultural Landscape: Understanding and Managing Change in Human-shaped Environments*, Plieninger, T., Bieling, C. (eds). Cambridge University Press, Cambridge.

Kinzig, A., Ryan, P., Etienne, M., Allison, H., Elmqvist, T., Walker, B. (2006). Resilience and regime shifts: Assessing cascading effects. *Ecology and Society*, 11(1) [Online]. Available at: https://www.ecologyandsociety.org/vol11/iss1/art20/ [Accessed 5 May 2020].

Kirchhoff, T., Brand, F.S., Hoheisel, D. (2012). From cultural landscapes to resilient social–ecological systems: Transformation of a classical paradigm or a novel approach? In *Resilience and the Cultural Landscape: Understanding and Managing Change in Human-shaped Environments*, Plieninger, T., Bieling, C. (eds). Cambridge University Press, Cambridge.

Klein, R.J.T., Smit, M.J., Goosen, H., Hulsbergen, C.H. (1998). Resilience and vulnerability: Coastal dynamics or Dutch dikes? *The Geographical Journal*, 164(3), 259–268.

Lacoste, Y. (2003). *De la géopolitique aux paysages : dictionnaire de la géographie*. Armand Colin, Paris.

Laffont, P.-Y. (2008). Châteaux, Incastellamento et réseau viaire au Moyen Âge en France méridionale. In *Marqueurs des paysages et systèmes socio-économiques : actes du Colloque COST-[Action 27] du Mans, 7–9 décembre 2006*, Compatangelo-Soussignan, R., Bertrand, J.R., Chapman, J., Laffont, P.-Y. (eds). Presses universitaires de Rennes, Rennes.

Larousse, P. (1922). *Larousse universel en deux volumes : nouveau dictionnaire encyclopédique*. Larousse, Paris.

Larousse, P. and Augé, C. (1898). *Nouveau Larousse illustré : dictionnaire universel*. Larousse, Paris.

Lassalle, T. (1990). *Cartographie : 4 000 ans d'aventure et de passion*. Nathan, Paris.

Latour, B. (1993). *We Have Never Been Modern*. Harvard University Press, Cambridge.

Latour, B. (1997). *Nous n'avons jamais été modernes : essai d'anthropologie symétrique*. La Découverte, Paris.

Lavedan, P. (1926a). *Qu'est-ce que l'urbanisme ? Introduction à l'histoire de l'urbanisme.* Henri Laurens, Paris.

Lavedan, P. (1926b). *Histoire de l'urbanisme, Antiquité, Moyen Âge.* Henri Laurens, Paris.

Lavedan, P. (1941). *Histoire de l'urbanisme, Renaissance et temps moderne.* Henri Laurens, Paris.

Lavedan, P. (1959). *Géographie des villes.* Gallimard, Paris.

Lavigne, C. (1997). Parcellaires de fondation et parcellaires de formation à l'époque médiévale en Gascogne, Clefs de lecture et problèmes d'interprétation. In *Les formes du paysage 3. L'analyse des systèmes spatiaux*, Chouquer, G. (ed.). Errance, Paris.

Lavigne, F. and Belizal, E.D. (2010). Les effets géographiques des éruptions volcaniques. *EchoGéo* [Online]. Available at: http://journals.openedition.org/echogeo/12226 [Accessed 12 June 2018].

Le Blanc, A. (2015). Resilience, memory and practices. In *Resilience Imperative: Uncertainty, Risks and Disasters*, Reghezza-Zitt, M. and Rufat, S. (eds). ISTE Press, London, and Elsevier, Oxford.

Le Blanc, A. (2018). Les territoires rassurants. HDR, Université du littoral-côte d'Opale, Dunkirk [Online]. Available at: https://hal.archives-ouvertes.fr/tel-01872489 [Accessed 1July 2019].

Le Corbusier (1957). *La Charte d'Athènes. Suivi d'entretien avec les étudiants des écoles d'architecture.* Éditions de Minuit, Paris.

Le Corbusier (1994). *Urbanisme*. Flammarion, Paris.

Le Corbusier, Eardley, A., Sert, J. L. (1973). *The Athens Charter*. Grossman Publishers, New York.

Le Lannou, M. (1978). Les paysages fossiles (conférence inaugurale). In *Actes du colloque "Archéologie du paysage", XIII, Tours, Caesarodunum*, Chevallier, R. (ed.). Centre A. Piganiol, Tours.

Le Moigne, J.-L. (1994). *La théorie du système général : théorie de la modélisation.* PUF, Paris.

Lebel, L., Anderies, J., Campbell, B., Folke, C., Hatfield-Dodds, S., Hughes, T., Wilson, J. (2006). Governance and the capacity to manage resilience in regional social-ecological systems. *Ecology and Society*, 11(1) [Online]. Available at: https://www.ecologyandsociety.org/vol11/iss1/art19/ [Accessed 13 June 2018].

Lefebvre, B. (2008). La formation d'un tissu urbain dans la cité de Tours : du site de l'amphithéâtre antique au quartier canonial (V[e]–XVIIIe s.). PhD Thesis, Université François Rabelais, Tours [Online]. Available at: https://tel.archives-ouvertes.fr/tel-00349580 [Accessed 5 July 2019].

Lefebvre, B. (2011). Modèles de réutilisation des amphithéâtres antiques dans la formation des tissus urbains. *M@ppemonde*, 101 [Online]. Available at: http://mappemonde.mgm.fr/num29/articles/art11102.html [Accessed 21 October 2016].

Lepetit, B. (1984). *Chemins de terre et voies d'eau : réseaux de transports et organisation de l'espace en France 1740–1840*. Éditions de l'École des hautes études en sciences sociales, Paris.

Lepetit, B. (1995). *Les formes de l'expérience*. Albin Michel, Paris.

Lepetit, B. and Pumain, D. (eds) (1993). *Temporalités urbaines*. Anthropos, Paris.

Leturcq, S. (1997). La route et le paysage. Dynamique et stabilité des réseaux routiers beaucerons entre Étampes et la forêt d'Orléans. In *Les Formes du paysage*, volume 3, Chouquer, G. (ed.). Errance, Paris.

Leturcq, S. (2008). Fonction et devenir d'un réseau invisible : les crêtes de labours dans les terroirs beaucerons (XIV^e^–XX^e^ siècles). In *Marqueurs des paysages et systèmes socio-économiques : actes du Colloque COST-[Action 27] du Mans, 7–9 décembre 2006*, Compatangelo-Soussignan, R., Bertrand, J.R., Chapman, J., Laffont, P.-Y. (eds). Presses universi-taires de Rennes, Rennes.

Leveau, P. (1997). Temps, espace et structuration des paysages. In *Les formes du paysage*, volume 3, Chouquer, G. (ed.). Errance, Paris.

Lévêque, C., Muxart, T., Abbadie, L., Weill, A., Van der Leeuw, S. (2003). L'anthroposystème : entité structurelle et fonctionnelle des interactions sociétés-milieux. In *Quelle nature voulons-nous ? Pour une approche socio-écologique du champ de l'environnement*, Lévêque, C., Van der Leeuw, S. (eds). Elsevier, Paris.

Lévy, J. and Lussault, M. (eds). (2003). *Dictionnaire de la géographie*. Belin, Paris.

Levy, T.E. (1983). The emergence of specialized pastoralism in the southern Levant. *World Archaeology*, 15(1), 15–36.

Lhomme, S. (2012). Les réseaux techniques comme vecteur de propagation des risques en milieu urbain. Une contribution théorique et pratique à l'analyse de la résilience urbaine. PhD thesis, Université Paris-Diderot – Paris VII, Paris [Online]. Available at: https://tel.archives-ouvertes.fr/tel-00772204/document [Accessed 1 April 2017].

Lhomme, S., Serre, D., Diab, Y., Laganier, R. (2010). Les réseaux techniques face aux inondations ou comment définir des indicateurs de performance de ces réseaux pour évaluer la résilience urbaine. *Bulletin de l'Association des géographes français*, 87(4), 487–502.

Liger, B. (1972). Le traitement de l'information dans l'espace historique. PhD Thesis, Université François Rabelais, Tours.

Lucas, G. (2012). *Understanding the Archaeological Record*. Cambridge University Press, Cambridge.

Maitland, F.W. (1987). *Domesday Book and Beyond: Three Essays in the Early History of England*. Cambridge University Press, Cambridge.

Marchand, C. (1997). Réseau viaire et dessin parcellaire : étude morphologique de la région du Gâtinais oriental. In *Les formes du paysage*, volume 3, Chouquer, G. (ed.). Errance, Paris.

Marchand, C. (2000). Recherches sur les réseaux de formes ; processus dynamiques des paysages du Sénonais occidental. PhD Thesis, Université François Rabelais, Tours.

Marchand, C. (2003). Des centuriations plus belles que jamais ? *Études rurales*, 167–168, 93–114.

Maret, I. and Cadoul, T. (2008). Résilience et reconstruction durable : que nous apprend La Nouvelle-Orléans ? Urban resiliency and sustainable rebuilding: What can we learn from New Orleans? *Annales de géographie*, 663, 104–124.

Martin-Breen, P. and Anderies, J.M. (2011). Resilience: A literature review. Report, Rockfeller Foundations, New York [Online]. Available at: https://opendocs.ids.ac.uk/opendocs/bitstream/handle/123456789/3692/Bellagio-Rockefeller%20bp.pdf?sequence=1 [Accessed 2 June 2016].

Masutti, C. (2002). L'héritage d'Henri Baulig (1877–1962) : éléments biographiques, bibliographie complète, inventaire des notes manuscrites. *Cybergeo: European Journal of Geography*, 22 [Online]. Available at: https://cybergeo.revues.org/3520 [Accessed 21 October 2016].

Maurin, A. (1992). *Le cadastre en France, histoire et rénovation*. CNRS, Paris.

Meerow, S. and Newell, J.P. (2016). Urban resilience for whom, what, when, where, and why? *Urban Geography*, 40(3), 1–21.

Meerow, S., Newell, J.P., Stults, M. (2016). Defining urban resilience: A review. *Landscape and Urban Planning*, 147, 38–49.

Meitzen, A. (1895). *Siedelung und agrarwesen der Westgermanen und Ostgermanen, der Kelten, Römer, Finnen und Slawen*. W. Hertz, Berlin.

Mennessier-Jouannet, C. and Buchsenschutz, O. (1996). Des "champs celtiques" aux systèmes de parcellaires protohistoriques. In *Les Formes du paysage, Tome 2, Archéologie des parcellaires*, Chouquer, G. (ed.). Errance, Paris.

Merleau-Ponty, M. (1979). *Le Visible et l'invisible : suivi de notes de travail*. Gallimard, Paris.

Meynier, A. (1969). *Histoire de la pensée géographique*. Presses universitaires de France, Paris.

Meynier, A. (1970). *Les paysages agraires*. Armand Colin, Paris.

Morin, E. (1977). *La méthode*. Le Seuil, Paris.

Mumby, P.J., Hastings, A., Edwards, H.J. (2007). Thresholds and the resilience of Caribbean coral reefs. *Nature*, 450(7166), 98–101.

Muxart, T., Vivien, F.-D., Villalba, B., Burnouf J. (eds). (2003). *Des milieux et des hommes : fragments d'histoires croisées*. Elsevier, Paris.

Nelson, M.C., Hegmon, M., Kulow, S.R., Schollmeyer, K.G. (2006). Archaeological and ecological perspectives on reorganization: A case study from the Mimbres region of the U.S. Southwest. *American Antiquity*, 71(3), 403–432.

Nelson, M.C., Hegmon, M., Kulow, S.R., Peeples, M.A., Kintigh, K.W., Kinzig, A.P. (2011). Resisting diversity: A long-term archaeological study. *Ecology and Society*, 16(1) [Online]. Available at: http://www.ecologyandsociety.org/vol16/iss1/art25/ [Accessed 16 April 2018].

Newman, P. and Jennings, I. (2008). *Cities as Sustainable Ecosystems: Principles and Practices.* Island Press, Washington.

Noizet, H. (2005). La transmission de la "nature" et du "rural" dans la ville : le cas de tours. *Études rurales*, 175–176, 109–128.

Noizet, H. and Clémençon, A.-S. (2020). *Faire ville. Entre planifié et impensé, la fabrique ordinaire des formes urbaines.* Presses universitaires de Vincennes, Vincennes.

Noizet, H., Robert, S., Mirlou, L. (2011). Cartographie des crues centennales à Paris (1740, 1910). *Revue du Nord*, 26, 91–104 [Online]. Available at: http://halshs.archives-ouvertes.fr/halshs-00668371 [Accessed 16 April 2018].

Noizet, H., Mirlou, L., Robert, S. (2013). La résilience des formes. La ceinture urbaine de Paris sur la rive droite. *Études rurales*, 191, 191–219 [Online]. Available at: 10.4000/etudesrurales.9851 [Accessed 16 April 2018].

Norgaard, R.B. (2009). *Development Betrayed: The End of Progress and a Coevolutionary Revisioning of the Future.* Routledge, London.

Nuninger, L. (2002). Peuplement et territoires protohistoriques du VIIIe au Ier s. av. J.-C. en Languedoc oriental (Gard-Hérault). PhD Thesis, Université de Franche-Comté, Besançon [Online]. Available at: https://tel.archives-ouvertes.fr/tel-00002981 [Accessed 5 July 2019].

Olivier, L. (2009). *Le sombre abîme du temps. Mémoire et archéologie.* Le Seuil, Paris.

Ostrom, E. (1990). *Governing the Commons: The Evolution of Institutions for Collective Action.* Cambridge University Press, Cambridge/New York.

Oxford University Press (1991). *The Oxford English Dictionary.* Oxford University Press, London.

OUP (2005). *Pocket Oxford Latin Dictionary*, Revised edition. OUP, Oxford.

Parent, M. (1948). Les civilisations vivantes. Les villes et leurs structures. In *La découverte aérienne du monde*, Chombart de Lauwe, P.H. (ed.). Horizons de France, Rennes.

Pérol, C. (2007). Comment penser la route ? *Siècles. Cahiers du Centre d'histoire "Espaces et Cultures"*, 25, 3–14.

Perring, M.P., Frenne, P.D., Baeten, L., Maes, S.L., Depauw, L., Blondeel, H., Carón, M.M., Verheyen, K. (2015). Global environmental change effects on ecosystems: The importance of land-use legacies. *Global Change Biology*, 22(4), 1361–1371.

Pierdet, C. (2009). Les temporalités de la relation ville-fleuve à Phnom-Penh (Cambodge) : la fixation d'une capitale fluviale par la construction d'un système hydraulique (1865–2005). PhD Thesis, Université Paris I, Panthéon-Sorbonne, Paris [Online]. Available at: http://crdaln.documentation.developpement-durable.gouv.fr/documents/Crdaln/0238/Cdu-0238604/Resume_Pierdet_C.pdf [Accessed 28 October 2015].

Pierdet, C. (2015). Resilience: A systemic property. *Resilience Imperative: Uncertainty, Risks and Disasters*. Reghezza-Zitt, M. and Rufat, S. (eds). ISTE Press, London, and Elsevier, Oxford.

Pigeon, P. (2012). Apports de la résilience à la géographie des risques : l'exemple de La Faute-sur-Mer (Vendée, France). *VertigO – La revue électronique en sciences de l'environnement*, 12(1) [Online]. Available at: http://vertigo.revues.org/12031 [Accessed 18 September 2017].

Pimm, S.L. (1984). The complexity and stability of ecosystems. *Nature*, 307(5949), 321–326.

Pinchemel, P. and Pinchemel, G. (1994). *La face de la terre : éléments de géographie*. Armand Colin, Paris.

Pinon, P. (1978). Le passage des structures architecturales antiques dans les tissus urbains médiévaux. *Caesarodunum*, 1978, 387–395.

Pinon, P. (1994). La lecture des persistances dans les formes urbaines et leur interpétation historique : le cas des villes d'origine romaine en Gaule. *Caesarodunum*, XXVIII(2), 39–49.

Pinoteau, C. (2003). Changer la carte, c'est changer l'objet. *Études rurales*, 167–168, 247–262.

Pinoteau, C. and Di Pietro, F. (2003). Association de formes et de dynamiques dans le bassin versant de l'Aubrière (Indre-et-Loire). *Études rurales*, 167–168, 263–284.

Piveteau, J.-L. (1990). L'épaisseur temporelle de l'organisation de l'espace : "palimpseste" et "coupe tranversale". *Géopoint*, 211–220.

Plieninger, T. and Bieling, C. (eds). (2012). *Resilience and the Cultural Landscape: Understanding and Managing Change in Human-shaped Environments*. Cambridge University Press, Cambridge.

Poëte, M. (1924). *Une vie de cité. Paris de sa naissance à nos jours. Tome 1. La jeunesse. Des origines aux temps modernes*. Picard, Paris.

Poëte, M. (2000). *Introduction à l'urbanisme*. In Bardet, G., Bardet, F., Tonka, H. (eds). Sens & Tonka, Paris.

Poidebard, A. (1934). *La trace de Rome dans le désert de Syrie, le limes de Trajan à la conquête Arabe, recherches aériennes (1925–1932)*. Haut-Commissariat de la République française en Syrie et Paul Geuthner, Paris.

Poidebard, A. and Mouterde, R. (1945). *Le Limes de Chalcis*. Geuthner, Paris.

Poïesis (1998). Dossier : La ville entre ordre et désordre. *Poïesis Architecture, arts, science et philosophie*, 7.

Poignault, R. (2007). Hommage au professeur Raymond Chevalier (1929–2004). In *Du mythe à la géopolitique : Europe entre Orient et Occident*, de Croizant, O.W., Montifroy, G.A. (eds). L'Âge d'Homme, Lausanne.

Poirier, N. (2007). Un espace rural en Berry dans la longue durée : expérience de micro-analyse des dynamiques spatio-temporelles du paysage et du peuplement dans la région de Sancergues (Cher). PhD Thesis, Université François Rabelais, Tours.

Prigogine, I. and Stengers, I. (1990). *La nouvelle alliance : métamorphose de la science.* Gallimard, Paris.

Pumain, D. (2003). Une approche de la complexité en géographie. *Géocarrefour*, 78(1), 25–31.

Pumain, D. (2004). Système, Hypergéo [Online]. Available at: http://www.hypergeo.eu/spip.php?article5 [Accessed 29 November 2018].

Pumain, D., Sanders, L., Saint-Julien, T. (1989). *Villes et auto-organisation.* Économica, Paris.

Quenault, B. (2013a). Du double affrontement ontologique/axiologique autour de la résilience aux risques de catastrophe : les spécificités de l'approche française. *VertigO*, 13(3) [Online]. Available at: http://vertigo.revues.org/14510 [Accessed 18 September 2017].

Quenault, B. (2013b). Retour critique sur la mobilisation du concept de résilience en lien avec l'adaptation des systèmes urbains au changement climatique. *EchoGéo* [Online]. Available at: https://echogeo.revues.org/13403 [Accessed 18 September 2017].

Quenault, B. (2015). De Hyōgo à Sendai, la résilience comme impératif d'adaptation aux risques de catastrophe : nouvelle valeur universelle ou gouvernement par la catastrophe ? *Développement durable et territoires. Économie, géographie, politique, droit, sociologie*, 6(3) [Online]. Available at: http://developpementdurable.revues.org/11010 [Accessed 18 September 2017].

Randall, H.J. (1934). History in the open air. *Antiquity*, 8(29), 5–23.

Raynaud, C. (2003). Les systèmes agraires antiques : quelle approche archéologique ? *Revue archéologique de Picardie*, 1(1), 281–298.

Redman, C.L. and Kinzig, A.P. (2003). Resilience of past landscapes: Resilience theory, society, and the Longue Durée. *Conservation Ecology*, 7(1), 14 [Online]. Available at: http://www.consecol.org/vol7/iss1/art14/.

Redman, C.L., Nelson, M.C., Kinzig, A.P. (2009). The resilience of socioecological landscapes. In *The Archaeology of Environmental Change: Socionatural Legacies of Degradation and Resilience*, Fisher, C.T., Hill, J.B., Feinman, G.M. (eds). University of Arizona Press, Tucson.

Reghezza-Zitt, M. (2013). Utiliser la polysémie de la résilience pour comprendre les différentes approches du risque et leur possible articulation. *EchoGéo* [Online]. Available at: https://journals.openedition.org/echogeo/13401 [Accessed 25 June 2019].

Reghezza-Zitt, M. (2015). De l'avènement du Monde à celui de la planète : le basculement de la société du risque à la société de l'incertitude. HDR thesis, Université de Paris I Panthéon-Sorbonne, Paris [Online]. Available at: https://tel.archives-ouvertes.fr/tel-0125 5031 [Accessed 25 June 2019].

Reghezza-Zitt, M. and Rufat, S. (eds) (2015). *Resilience Imperative: Uncertainty, Risks and Disasters*. ISTE Press, London, and Elsevier, Oxford.

Reghezza-Zitt, M., Rufat, S., Djament-Tran, G., Le Blanc, A., Lhomme, S. (2012). What resilience is not: Uses and abuses. *Cybergeo: European Journal of Geography*, 18 [Online]. Available at: http://journals.openedition.org/cybergeo/25554 [Accessed 5 December 2019].

Renfrew, C. (1978). Trajectory discontinuity and morphogenesis: The implications of catastrophe theory for archaeology. *American Antiquity*, 43(2), 203–222.

Renouvier, C. (1876). *Uchronie (l'utopie dans l'histoire) : esquisse historique apocryphe du développement de la civilisation européenne tel qu'il n'a pas été, tel qu'il aurait pu être.* La Critique philosophique, Paris.

Rey, A. (1993). *Dictionnaire historique de l'ancien langage français.* Dictionnaire Le Robert, Paris.

Rey, A. and Rey-Debove, J. (eds). (1992). *Le Petit Robert. Dictionnaire alphabétique et analogique de la langue française.* Le Robert, Paris.

Rhemtulla, J.M. and Mladenoff, D.J. (2007). Why history matters in landscape ecology? *Landscape Ecology*, 22(1), 1–3.

Rice, P.M. (2013). Time, memory, and resilience among the Maya. In *Millenary Maya Societies: Past Crises and Resilience*, Arnauld, C., Breton, A. (eds). Mesoweb, 10–21 [Online]. Available at: http://www.mesoweb.com/publications/MMS/1_Rice.pdf [Accessed 5 January 2019].

Robert, S. (1994). Étude chrono-morphologique du paysage sénartais. In *Rapport de synthèse du Projet collectif interrégional "Cartographie des anciens parcellaires de la France"*, Chouquer, G. (ed.). CNRS/ Université François Rabelais, Tours.

Robert, S. (1996a). Étude morphologique du parcellaire et de la voirie de la région de Beaugency. In *Les Formes du paysage*, volume 1, Chouquer, G. (ed.). Errance, Paris.

Robert, S. (1996b). Le Parcellaire du plateau de Sénart (Seine-et-Marne). In *Les Formes du paysage*, volume 1, Chouquer, G. (ed.). Errance, Paris.

Robert, S. (1997). Le relevé parcellaire sur photographies aériennes et cartes anciennes. In *Les Formes du paysage*, volume 3, Chouquer, G. (ed.). Errance, Paris.

Robert, S. (2003a). Comment les formes du passé se transmettent-elles ? *Études rurales*, 167–168, 115–132 [Online]. Available at: https://etudesrurales.revues.org/2938 [Accessed 20 July 2019].

Robert, S. (2003b). L'analyse morphologique entre archéologie, urbanisme et aménagement du territoire. Exemples d'études de formes urbaines et rurales dans le Val-d'Oise. PhD Thesis, Université Paris I Panthéon-Sorbonne, Paris [Online]. Available at: http://tel.archives-ouvertes.fr/tel-00371063 [Accessed 20 September 2013].

Robert, S. (2003c). Archéologie préventive et morphologie : deux points de vues scientifiques différents. In *Actes du colloque AGER V : Actualité de la recherche en histoire et archéologie agraires, Besançon, 19–20 septembre 2000*. Presses universitaires de Franche-Comté, Besançon, 41–53 [Online]. Available at: http://hal.archives-ouvertes.fr/hal-00371112 [Accessed 20 September 2018].

Robert, S. (2009). De la route-monument au réseau routier. *Les Nouvelles de l'archéologie*, 115, 8–12 [Online]. Available at: https://doi.org/10.4000/nda.631 [Accessed 20 September 2013].

Robert, S. (ed.). (2011). *Sources et techniques de l'archéogéographie*. Presses universitaires de Franche-Comté, Besançon.

Robert, S. (ed.). (2012a). Dossier : Archéologie et géographie. *L'Espace géographique*, 41(4), 289–361 [Online]. Available at: https://www.cairn.info/revue-espace-geographique-2012-4.htm [Accessed 20 June 2019].

Robert, S. (2012b). Une vision renouvelée de la dynamique forme-société entre archéologie et géographie. *L'Espace géographique*, 41(4), 310–323 [Online]. Available at: https://www.cairn.info/revue-espace-geographique-2012-4.htm [Accessed 20 June 2019].

Robert, S. (2013). Des outils pour l'analyse de la transmission des trames et réseaux planimétriques : approches multiscalaires et traitements sur SIG. In *Paysage et environnement, de la reconstitution du passé aux modèles pro-spectifs, colloque de Chilhac 27–30 sept. 2006*, Galop, D. (ed.). Presses universi-taires de Franche-Comté, Besançon, 25–39 [Online]. Available at: http://halshs.archives-ouvertes.fr/halshs-00409104 [Accessed 20 September 2013].

Robert, S. (2017). Archéologie du paysage et géographie : entre observation, transferts et co-constructions. In *P@lethnologie, Archéologie et sciences sociales*, Boissinot, P. (ed.). Université Toulouse Jean Jaurès, Toulouse [Online]. Available at: http://blogs.univ-tlse2.fr/palethnologie/2017-05-robert/ [Accessed 28 June 2018].

Robert, S. (2018). Les chasse-marée et la route du poisson. Réduire la distance pour approvisionner Paris du XIIIe siècle au XVIIIe siècle. *Techniques & Culture. Revue semestrielle d'anthropologie des techniques*, 69 [Online]. Available at: http://journals.openedition.org/tc/8931 [Accessed 26 July 2018].

Robert, S. (2020). Résilience, changement, persistance. In *Le temps long du peuplement : concepts et mots-clés*, Sanders, L., Bretagnolle, A., Brun, P., Ozouf-Marignier, M.-V., Verdier, N. (eds.) Presses universitaires François Rabelais, Tours.

Robert, S. (2021). Vers un nouveau modèle analytique en archéologie routière. *Actes du colloque Voies, réseaux, paysages en Gaule, en hommage à Jean-Luc Fiches, UMR 5140 Archéologies des sociétés méditerranéennes – Revue archéologique de Narbonnaise*, Raynaud, C. (ed.), 49, 477–491.

Robert, S. and Costa, L. (2014). Chapitre premier : Analyse des réseaux géographiques du Término de Villajoyosa. In *Villajoyosa antique (Alicante, Espagne), Terres et topographie, le sanctuaire de la Malladeta*, Rouillard, P., Espinosa, A., Moratalla, J. (eds). Casa Velasquez, Madrid.

Robert, S. and Sittler, B. (eds). (2016), Water as a morphogen in landscapes – L'eau comme morphogène dans les paysages. In *Proceedings of the XVII UISPP World Congress*, 1–7 September 2014, (4)/Session A14, 104, Archaeopress, Oxford, Burgos [Online]. Available at: http://www.archaeopress.com/ArchaeopressShop/Public/download.asp?id=%7BD195E698-1FFB-46E4-BFBD-F7E7F6A5A9FF%7D [Accessed 20 September 2013].

Robert, S. and Verdier, N. (eds). (2009). Une archéologie des réseaux viaires. *Les Nouvelles de l'archéologie*, 115, 5–57 [Online]. Available at: https://doi.org/10.4000/nda.631 [Accessed 20 September 2013].

Robert, S. and Verdier, N. (eds). (2014). *Dynamique et résilience des réseaux routiers. Archéogeographes et archéologues en région Île-de-France*. FERACF, Tours.

Robert, S., Noizet, H., Grosso, E., Chareille, P. (2013). Analyses morphologiques du parcellaire ancien de Paris. In *Paris, de parcelles en pixels : analyse géomatique de l'espace parisien médiéval et moderne*, Noizet, H., Bove, B., Costa, L. (eds). Presses universitaires de Vincennes, Vincennes.

Robic, M.-C. (2012). Du ciel au sol : la vue aérienne et l'idéal de la vue raisonnée. Les photographies aériennes dans la géographie des années 1920. In *Vues aériennes. Seize études pour une histoire culturelle*, Pousin, F., Dorian, M. (eds). Métis-Presses, Geneva.

Rodier, X. (2016). Espace-temps des sociétés du passé. HDR, Université de Franche-Comté, Besançon [Online]. Available at: https://halshs.archives-ouvertes.fr/tel-01259368 [Accessed 24 June 2019].

Rossi, A. (2001). *L'architecture de la ville*. Infolio, Dijon.

Rufat, S. (2012). Existe-t-il une "mauvaise" résilience ? In *Résiliences urbaines : les villes face aux catastrophes*, Djament-Tran, G., Reghezza-Zitt, M. (eds). Le Manuscrit, Paris.

de Ruig, J.H.M. (1997). Resilience in Dutch coastline management. *Coastline*, 6(2), 4–8.

Sanders, L. (1992). *Système de villes et synergétique*. Anthropos, Paris.

Sanders, L. (2007). *Models in Spatial Analysis*. ISTE, London.

Sanders, L. (ed.) (2017). *Peupler la Terre. De la préhistoire à l'ère des métropoles*. Presses universitaires François Rabelais, Tours [Online]. Available at: https://books.openedition.org/pufr/10467.

Sanders, L. (2020) Émergence, transitions et bifurcations. In *Le temps long du peuplement : concepts et mots-clés*, Sanders, L., Bretagnolle, A., Brun, P., Ozouf-Marignier, M.-V., Verdier, N. (eds). Presses universitaires François Rabelais, Tours.

Sanders, L., Brun, P., Robert, S. (2020a). Émergence, transmission et résilience, trois processus clés de l'évolution des systèmes de peuplement. In *Le temps long du peuplement : concepts et mots-clés*, Sanders, L., Bretagnolle, A., Brun, P., Ozouf-Marignier, M.-V., Verdier, N. (eds). Presses universitaires François Rabelais, Tours.

Sanders, L., Bretagnolle, A., Brun, P., Ozouf-Marignier, M.-V., Verdier, N. (eds) (2020b). *Le temps long du peuplement : concepts et mots-clés*. Presses universitaires François Rabelais, Tours.

Sauer, J.D. (1977). Biogeographical theory and cultural analogies. *World Archaeology*, 8(3), 320–331.

Sauer, C.O. and Leighly, J. (1963). *Land and Life: A Selection from the Writings of Carl Ortwin Sauer*. University of California Press, Berkeley.

de Saussure, F. (1995a). *Cours de linguistique générale*. Payot, Paris.

de Saussure, F. (1995b). *Course in General Linguistics*. McGraw-Hill, New York.

Scheffer, M. and Carpenter, S.R. (2003). Catastrophic regime shifts in ecosystems: Linking theory to observation. *Trends Ecol. Evol.*, 18(12), 648–656.

Scheffer, M., Hosper, S., Meijer, M., Moss, B., Jeppesen, E. (1993). Alternative equilibria in shallow lakes. *Trends in Ecology & Evolution*, 8(8), 275–279.

Scheffer, M., Carpenter, S., Foley, J.A., Folke, C., Walker, B. (2001). Catastrophic shifts in ecosystems. *Nature*, 413(6856), 591–596.

Scheffer, M., Westley, F., Brock, W., Holmgren, M. (2002). Dynamic interactions of societies and ecosystems linking theories from ecology, economy, and sociology. In *Panarchy: Understanding Transformations in Human and Natural Systems*, Gunderson, L.H., Holling, C.S. (eds). Island Press, Washington.

Scheffer, M., Carpenter, S.R., Lenton, T.M., Bascompte, J., Brock, W., Dakos, V., van de Koppel, J., van de Leemput, I.A., Levin, S.A., van Nes, E.H., Pascual, M., Vandermeer, J. (2012). Anticipating critical transitions. *Science*, 338(6105), 344–348.

Schoon, M. and Fabricius, C. (eds) (2011). Ecology and society: Long-term vulnerability and transformation. *Ecology and Society*, 16(2) [Online]. Available at: https://www.ecologyandsociety.org/issues/view.php/feature/46 [Accessed 16 April 2018].

Schwartz, G.M. and Nichols, J.J. (eds) (2006). *After Collapse: The Regeneration of Complex Societies*. University of Arizona Press, Tucson.

Scollar, I., Tabbagh, A., Hesse, A., Herzog, I. (1990). *Archaeological Prospecting and Remote Sensing*. Cambridge University Press, Cambridge.

Serre, D. (2011). La ville résiliente aux inondations. Méthodes et outils d'évaluation. PhD Thesis, Université Paris-Est, Marne-la-Vallée [Online]. Available at: https://tel.archives-ouvertes.fr/tel-00777206 [Accessed 18 April 2019].

Sitte, C. (1890). *L'art de bâtir les villes : notes et réflexions d'un architecte/Camillo Sitte ; traduites et complétées par Camille Martin.* Henri Laurens, Paris.

Soulié, V. and Cyrulnik, B., (2000). *La résilience : un nouveau concept psychologique.* Arts et éducation, Paris.

Steinberg, J. (1982). *La carte topographique.* SEDES, Paris.

Stübben, J. (1890). *Der Städtebau.* J.M. Gebhardt, Leipzig.

Szabó, P. (2014). Historical ecology: Past, present and future. *Biological Reviews*, 90(4), 997–1014.

Taillet, R., Villain, L., Febvre, P. (2013). *Dictionnaire de physique.* De Boeck Supérieur, Louvain-la-Neuve.

Tainter, J.A. (1988). *The Collapse of Complex Societies.* Cambridge University Press, London.

Thom, R. (1973). *Stabilité structurelle et morphogenèse.* W.A. Benjamin Verlag, New York.

Thompson, V.D. and Turck, J.A. (2009). Adaptive cycles of coastal hunter-gatherers. *American Antiquity*, 74(2), 255–278.

Timmerman, P. (1981). Vulnerability, resilience and the collapse of society. A review of models and possible climatic applications. Institute for Environmental Studies, University of Toronto, Toronto [Online]. Available at: http://www.ilankelman.org/miscellany/Timmerman1981.pdf.

Tisseron, S. (2007). *La résilience.* PUF, Paris.

Toubin, M., Lhomme, S., Diab, Y., Serre, D., Laganier, R. (2012). La Résilience urbaine : un nouveau concept opérationnel vecteur de durabilité urbaine ? *Développement durable et territoires. Économie, géographie, politique, droit, sociologie*, 3(1) [Online]. Available at: http://developpementdurable.revues.org/9208 [Accessed 1 April 2017].

Turner, M.D. (2014). Political ecology I: An alliance with resilience? *Progress in Human Geography*, 38(4), 616–623.

Unwin, R. (1981). *L'étude pratique des plans de villes.* L'Équerre, Paris.

Vale, L.J. and Campanella, T.J. (2005). *The Resilient City: How Modern Cities Recover from Disaster.* Oxford University Press, Oxford.

Van der Leeuw, S. (ed.) (1998). *Understanding the Natural and Anthropogenic Causes of Land Degradation and Desertification in the Mediterranean Basin: The Archaeomedes Project.* Office for the Official Publications of the European Communities, Luxembourg.

Van der Leeuw, S. (2006). Crises vécues, crises perçues. In *Temps et espaces des crises de l'environnement*, Beck, C., Luginbühl, Y., Muxart, T. (eds). Quae, Versailles.

Van der Leeuw, S. (2009). What is an "environmental crisis" to an archaeologist? In *The Archaeology of Environmental Change: Socionatural Legacies of Degradation and Resilience*, Fisher, C.T., Hill, J.B., Feinman, G.M. (eds). University of Arizona Press, Tucson.

Van der Leeuw, S. (2020). *Social Sustainability, Past and Future. Undoing Unintended Consequences for the Earth's Survival.* Cambridge University Press, Cambridge [Online]. Available at: https://www.cambridge.org/core/books/social-sustainability-past-and-future/811395DC3A8D82EAD39C45657B2FD1AD [Accessed 10 January 2020].

Van der Leeuw, S. and Aschan-Leygonie, C. (2001). A long-term perspective on resilience in socio-natural systems. In *Paper Presented at the Workshop on "System Shocks – System Resilience" Held in Abisko*, May 22–26, 2000, Santa Fe Institute working paper 2001-08-04 [Online]. Available at: https://sfi-edu.s3.amazonaws.com/sfi-edu/production/uploads/sfi-com/dev/uploads/filer/4b/b6/4bb68ae-0b24-4866-8fb2-47e8b92e90c1/01-08-042.pdf [Accessed 10 December 2020].

Van der Leeuw, S. and Redman, C.L. (2002). Placing archaeology at the center of socio-natural studies. *American Antiquity*, 67(4), 597–605.

Van der Leeuw, S. and Thiébault, S. (eds) (2000). Environnement et archéologie. *Les Nouvelles de l'archéologie*, 78, 5–35.

Vanistendael, S. (2002). *La résilience : un regard qui fait vivre.* Bayard, Paris.

Verdier, N. and Bretagnolle, A. (2005). Images d'un réseau en évolution : les routes de poste dans la France préindustrielle (XVIIe siècle–début XIXe siècle). *M@ppemonde*, 79(3), 8.

Verhulst, A.E. (1995). *Le paysage rural : les structures parcellaires de l'Europe du Nord-Ouest.* Brepols, Turnhout.

Vidal de La Blache, P. (1926). *Principles of Human Geography.* Constable, London.

Vidal de La Blache, P. (2015). *Principes de géographie humaine.* ENS, Lyon [Online]. Available at: http://books.openedition.org/enseditions/341.

Vion, E. (1989). L'analyse archéologique des réseaux routiers : une rupture méthodologique, des réponses nouvelles. *Paysages découverts : histoire, géographie et archéologie du territoire en Suisse romande*, 67–99.

Vue, B. (2002). Mauvais chemins et servitudes de passage. *Cahiers Haut Marnais, actes des journées d'Art et d'Histoire 1998*, 45–68.

Walker, B. and Salt, D. (2006). *Resilience Thinking: Sustaining Ecosystems and People in a Changing World.* Island Press, Washington.

Walker, B., Holling, C.S., Carpenter, S., Kinzig, A. (2004). Resilience, adaptability and transformability in social-ecological systems. *Ecology and Society*, 9(2) [Online]. Available at: https://www.ecologyandsociety.org/vol9/iss2/art5/ [Accessed 28 June 2018].

Walker, B., Gunderson, L., Kinzig, A., Folke, C., Carpenter, S., Schultz, L. (2006). A handful of heuristics and some propositions for understanding resilience in social-ecological systems. *Ecology and Society*, 11(1) [Online]. Available at: https://www.ecologyand society.org/vol11/iss1/art13/ [Accessed 4 October 2018].

Watteaux, M. (2009). La dynamique de la planimétrie parcellaire et des réseaux routiers en Vendée méridionale. Études historiographiques et recherches archéogéographiques. PhD Thesis, Université Panthéon-Sorbonne – Paris I, Paris [Online]. Available at: http://tel.archives-ouvertes.fr/tel-00421955 [Accessed 22 May 2019].

Watteaux, M. (2011). Dossier : L'archéogéographie : un état des lieux et de leurs dynamiques. *Les Nouvelles de l'archéologie* [Online]. Available at: http://journals.openedition.org/nda/1491 [Accessed 22 May 2019].

Weiberg, E. (2012). What can resilience theory do for (Aegean) archaeology? In *Matters of Scale: Processes and Courses of Events in Archaeology and Cultural History*, Burström, N.M., Fahlander, F. (eds). Stockholm University, Stockholm.

Weichselgartner, J. and Kelman, I. (2015). Geographies of resilience: Challenges and opportunities of a descriptive concept. *Progress in Human Geography*, 39(3), 249–267.

Weishampel, J.F., Chase, A.F., Chase, D.Z., Hightower, J.N. (2013). Remote sensing of below-canopy land use features from the Maya Polity of Caracol. In *Understanding Landscapes: From Land Discovery to Spatial Organization, Proceedings of the Sessions C19 and C22 of the XVI World Congress*, Djindjian, F., Robert, S. (eds). Florianópolis, September 4–10, 2011, British Archeological Report, Oxford.

Widgren, M. (2012). Resilience thinking versus political ecology: Understanding the dynamics of small-scale, labour intensive farming landscape. In *Resilience and the Cultural Landscape: Understanding and Managing Change in Human-shaped Environments*, Plieninger, T., Bieling, C. (eds). Cambridge University Press, Cambridge.

Yellen, J.E. (1977). Long term hunter-gatherer adaptation to desert environments: A biogeographical perspective. *World Archaeology*, 8(3), 262–274.

Yesner, D.R. (2008). Ecology in archaeology. In *Handbook of Archaeological Theories*, Bentley, R.A., Maschner, H.D.G., Chippindale, C. (eds). Rowman/Littlefield, New York.

Yesner, D.R., Ayres, W.S., Carlson, D.L., Davis, R.S., Dewar, R., Hassan, F.A., Hayden, B., Lischka, J.J., Sheets, P.D., Osborn, A., Pokotylo, D.L., Rogers, T., Spanier, E., Turner, B.L., Wreschner, E.E. (1980). Maritime hunter-gatherers: Ecology and prehistory. *Current Anthropology*, 21(6), 727–750.

Zimmerer, K.S. (1994). Human geography and the "new ecology": The prospect and promise of integration. *Annals of the Association of American Geographers*, 84, 108–125.

Ziter, C., Graves, R.A., Turner, M.G. (2017). How do land-use legacies affect ecosystem services in United States cultural landscapes? *Landscape Ecology*, 21(3), 1129–1142.

# Index

## D



## E



## F, G



## H



## I

## R



## S



## T

## U



## V

Other titles from

ISTE

in

Interdisciplinarity, Science and Humanities

## 2020

ATCHOUA Julien, BOGUL Jean-Jacques, DIALLO Saikou
*Digital Technologies and African Societies: Challenges and Opportunities (Africa Digital Age Set – Volume 1)*

CHERICI Céline
*From Clouds to the Brain: The Movement of Electricity in Medical Science*

GUILLAUME Astrid, KURTS-WÖSTE Lia
*Making Sense, Making Science*

SCHEPS Ruth, MAUREL Marie-Christine
*Mathematics in the Visual Arts*

PAVIE Xavier
*Critical Philosophy of Innovation and the Innovator (Innovation and Responsibility Set – Volume 6)*

## 2019

ANCORI Bernard
*The Carousel of Time: Theory of Knowledge and Acceleration of Time*

BOULEAU Gabrielle
*Politicization of Ecological Issues: From Environmental Forms to Environmental Motives*

GAYON Jean†, interviewed by PETIT Victor
*Knowledge of Life Today: Conversations on Biology*

KHATCHATOUROV Armen, with the collaboration of CHARDEL Pierre-Antoine, FEENBERG Andrew, PÉRIÈS Gabriel
*Digital Identities in Tension: Between Autonomy and Control*
*(Innovation and Responsibility Set – Volume 5)*

LENOIR Virgil Cristian
*Ethically Structured Processes*
*(Innovation and Responsibility Set – Volume 4)*

NIKOLOVA Blagovesta
*The RRI Challenge: Responsibilization in a State of Tension with Market Regulation*
*(Innovation and Responsibility Set – Volume 3)*

ZEMBRI-MARY Geneviève
*Project Risks: Actions Around Uncertainty in Urban Planning and Infrastructure Development*

## 2018

LOISY Catherine, COULET Jean-Claude
*Competence and Program-based Approach in Training: Tools for Developing Responsible Activities*
*(Innovation and Responsibility Set – Volume 1)*

PANSERA Mario, OWEN Richard
*Innovation and Development: The Politics at the Bottom of the Pyramid*
*(Innovation and Responsibility Set – Volume 2)*

## 2017

MAESSCHALCK Marc
*Reflexive Governance for Research and Innovative Knowledge*
*(Responsible Research and Innovation Set – Volume 6)*

PELLÉ Sophie
*Business, Innovation and Responsibility*
*(Responsible Research and Innovation Set – Volume 7)*

## 2016

GIANNI Robert
*Responsibility and Freedom: The Ethical Realm of RRI*
*(Responsible Research and Innovation Set – Volume 2)*

GRUNWALD Armin
*The Hermeneutic Side of Responsible Research and Innovation*
*(Responsible Research and Innovation Set – Volume 5)*

LENOIR Virgil Cristian
*Ethical Efficiency: Responsibility and Contingency*
*(Responsible Research and Innovation Set – Volume 1)*

PELLÉ Sophie, REBER Bernard
*From Ethical Review to Responsible Research and Innovation*
*(Responsible Research and Innovation Set – Volume 3)*

REBER Bernard
*Precautionary Principle, Pluralism and Deliberation: Science and Ethics*
*(Responsible Research and Innovation Set – Volume 4)*

Printed and bound by CPI Group (UK) Ltd, Croydon, CR0 4YY

15/09/2026

14962360-0004